Becoming

a

U.S. Marine

Adapting Basic Training
to the 21st Century

D1227396

Becoming
a
U.S. Marine

Adapting Basic Training to the 21st Century

Nathan Decety

Anchor Book Press · Palatine

Becoming a U.S. Marine:
        Adapting Basic Training to the 21st Century
Copyright © Nathan Decety 2019
Anchor Book Press, Ltd.
440 W Colfax Street, Unit 1132, Palatine, IL 60078
ISBN: 9781949109191
Printed in the United States

Acknowledgement

To all Marines, past and present.
Thank you for your sacrifices.

Table of Contents

Introduction

Most new Marines return home after Boot Camp to the foggy familiarity of their homes. There, friends from a seemingly distant past inevitably ask, "How was Boot Camp?" After trying to summarize my experiences a few times, I realized it was going to be too great a task to discuss with so many friends and acquaintances. Writing about it seemed to be an apt substitute. I began transcribing my notes before leaving for the second portion of basic training, Marine Combat Training (MCT). At this month-long training school, the project increased in scope.

By the end of the four-month basic cycle (the combination of Boot Camp and MCT), I had become disenchanted with the limitations of training. It appeared to me that we had wasted so much time learning a fraction of what we could have, and that Recruits were simply not pushed hard enough. I saw many ways in which the Marine Corps could improve with little to no cost. Much of the inefficacy and inadequacy could be reorganized away. After I vocalized my criticisms to peers and Noncommissioned Officers, I began to feel as though I was not alone in thinking this way, and that my recommendations could indeed be implemented to improve the Corps.

This book is split into two parts. The first consists of my broad reflections on Marine basic training in the context of our modern world. I include in this section my opinions on both Boot Camp and MCT. I propose a few modifications to Boot Camp, and a massive overhaul of MCT. The second is an account of my time in Marine Corps Boot Camp. Those who seek a direct first-hand experience of Boot Camp and a background on terms before reading about possible modifications are thus encouraged to begin with Part 2. As far as I know, this is the only written day-by-day account of Boot Camp in existence.

My personal experiences from the transcribed notes became a sort of primary source, a foil upon which to propose changes. While my experiences can be read as a standalone account of Boot Camp, I suggest it be seen as a somewhat entertaining firsthand account of a process that I would like to see amended. I hope this book may cause some reflection in the taxpayers and Officers who finance and lead, respectively, our operations and thereby ameliorate the Marine Corps.

Personal Background

The majority of Marines go through Boot Camp very soon after completing high school. It is a formative experience for these individuals, and it changes their character enormously, as suggested by the adage 'once a Marine, always a Marine.' I did not have the same relationship with the Marine Corps. Two years before - in college - I attended Marine Officer Candidate School (OCS). I did not complete the program and therefore did not commission as a Second Lieutenant (an Officer). The culture was just too different, I did not fit in, I stuck out too much.

Perhaps I was immature then. I was certainly inexperienced; it was a completely miserable shock for me. Further, my commander, a Captain, talked about how little he saw his family and friends as he flew above the skies of the Middle East. There was no way I planned to spend my 20's in a job that made me depressed without seeing those I cared about or finding any joy in my occupation. I thought I was done with the military forever after OCS. Even though I had the right to return, I did not.

For readers not familiar with military hierarchy, there are two sets of ranks in the Marine Corps: Officer and Enlisted. Officers are college graduates. Sans connection to their studies in college, their military training consists of 3 months of OCS, 6 months of "Basic School," followed by their MOS school (Military Occupational Specialty – the actual job one does in the military). Enlisted are most often high school graduates without a college degree. They make up the vast majority of the Corps. Their training consists of 3 months of Boot Camp and 1 month of "Marine Combat Training," followed by their MOS school.

Other than specific MOS such as pilots, Officers develop policies and procedures and are ultimately responsible for accomplishing whatever mission is assigned to their unit. It is typically a managerial role, a world of paperwork and signatures. Enlisted in the lower echelons of the Marine Corps accomplish tasks, in which they become experts. For instance, radio operators know everything about their radio and wave propagation, supply Marines are experts at the inventorying and processing of gear, low altitude air defense Marines can identify any known aerial vehicle and how to take it down. In the higher echelons of the Enlisted are the individuals who ensure that work is completed, these are Noncommissioned Officers (NCOs). NCOs are Enlisted who have been promoted; they train, manage, lead, and take care of their Marines. They are the backbone of the Corps and are some of the most incredible individuals in the world. NCOs often occupy vastly different roles: teachers, workers, physical fitness instructors, dietitians, counselors, leaders, and supervisors. A commissioned Officer is ultimately in command, assisted by an Enlisted member to advise him or her. This relationship is

3

necessary for a variety of reasons, but in general it improves the likelihood that the Officer develops informed policies; the Enlisted advisor has been doing a specific job for a long time and knows the Marines in the unit as well as their roles as part of that unit, whereas Officers change frequently and may not know either their subordinates or even all the necessary tasks needed to accomplished mission.

The divisions make more sense once numbers are applied. A full-sized infantry platoon includes about 40 Marines. In charge is a First or Second Lieutenant. The Lieutenant is typically assisted by a Staff Sergeant. The platoon is divided into three squads of 12-13, each led by a Sergeant. The squad is further divided into fire teams of 4, led by a Lance Corporal or Corporal. The structure is such that each rank only needs to manage about 4 people; the fire team leader manages their fire team, the squad leader manages the fire team leaders, the Staff Sergeant and Lieutenant manage the squad leaders; the Lieutenant holds the responsibility for the general welfare of the platoon and ensuring that the platoon's mission is accomplished. Such a management structure evolved directly from 20th century combat. The military realized that it was very difficult to effectively manage more than 4 or 5 people in modern warfare.

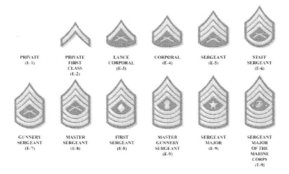

The Marine Corps Enlisted rank structure.

Corporals and above are Noncommissioned Officers, Staff Sergeants and above are called staff Noncommissioned Officers. The majority of the Marine Corps is composed of Privates, Private First Class, and Lance Corporals, who are all known as junior Marines. Promotion through to Lance Corporal is automatic.

The Marine Corps Officer rank structure.

To return from this tangent, I graduated from college two years after I attended OCS. I studied history and finance in school and loved those subjects. Naturally, I asked myself what I should do next. I decided to investigate what the older generation thought and asked what they would do in my shoes. Having agglomerated their opinions, I resolved that the best course of action was to learn and experience as much as possible, enjoy myself, and 'build my human capital,' by trying my hand at various jobs.

This of course brought me back to the idea of the Marine Corps. I was drawn to the warrior culture, to the ethos and difficulty of the Marine Corps. Based on the image conveyed by the Corps, the Corps is the elite fighting force of the United States. At the same time, I met innumerable Marines who told me camaraderie and friendship existed more on the Enlisted side, and specific programs existed with limited service obligations. That decided it for me. Over a few short months of training, I would earn a title, experience something interesting, serve my country, and hopefully make new lifelong friends. Remembering my experience from OCS, I knew I might forget much due to sleep deprivation, and that there would no doubt be important moments I would like to recall. I shipped out armed with notecards and a pen to create a Boot Camp journal. That journal was the impetus for the writing of this book. The entire Part 2 of this book is composed of the edited and transcribed notes that I maintained during Boot Camp.

Figure 1: My notes, I carried pieces of paper in my pockets and would write down my notes at opportune moments - usually to the derision of my fellow Recruits who obstinately thought I was writing poetry.

Every good Marine knows their leadership traits and principles. They will be brought up again several times in the coming discussion.

Marine Corps Leadership Traits	Marine Corps Leadership Principles
• Justice • Judgment • Dependability • Initiative • Decisiveness • Tact • Integrity • Endurance • Bearing • Unselfishness • Courage • Knowledge • Loyalty • Enthusiasm	• Know yourself and seek self-improvement • Be technically and tactically proficient • Develop a sense of responsibility among your subordinates • Make sound and timely decisions • Set the example • Know your Marines and look out for their welfare • Keep your Marines informed • Seek responsibility and take responsibility for your actions • Ensure assigned tasks are understood, supervised, and accomplished • Train your Marines as a team • Employ your team in accordance with its capabilities

Part 1

Nathan Decety

Proposed Modifications

The first part of this book is a more serious and original section than the second part. I argue for a major organizational pivot for the Marine Corps. I then discuss several important changes to a Marine Corps training schedules and methods which, I believe, would ameliorate the output Marine. The underlying thesis is that the Marine Corps is too large, not selective enough, and that leadership at the lowest level is a mirage which should be made into a reality at all phases of training.

The practical sources of this section are my personal experiences, incorporating opinions and feedback on operational procedures I have gathered over my time in the Marine Corps. The theoretical basis of this section are the works on the commandant's reading list – a set of books hand selected by the Commandant of the Marine Corps for the lessons they provide.[1] Following this set of data has two advantages. First, the reading list can be interpreted as the opinions and views of the upper echelon of Marine Corps leadership. With all their experience and knowledge, successive commandants and their advisers have decided that these specific works are highly relevant to combat and professional military practices. Second, the list

---

[1] I used the following from the commandant's reading list: Robert Neller, "2017 Marine Corps Commandant's Professional Reading List," Marine Corps University Research Library, May 16, 2017 (Accessed June 2, 2018) https://grc-usmcu.libguides.com/usmc-reading-list; see also, USMC Officer, "Commandant's Reading List – A Complete List," (Accessed June 2, 2018) https://www.usmcOfficer.com/marine-corps-knowledge/commandants-reading-list/.

constrains the possible sources of information. Evidence can always be cherry-picked to fit one's arguments, but if the source of that evidence is highly limited, it reduces the possibility of an unfounded argument on my part, so long as the sources are apt. The views and opinions expressed herein are my own and do not reflect those of the Marine Corps. The Marine Corps in no way endorsed or verified any statements made in this book.

Chapter 1: Size and Selectivity

The Marine Corps is too large. This statement may appear injudicious because the Marine Corps is the smallest branch of the U.S. armed forces and prides itself on doing more with less: more missions and more training, with worse equipment and fewer individuals. I shall nevertheless argue this point by looking at macro trends in the United States population. I will then discuss how the relationship between the Marine Corps and the Navy can be leveraged to ameliorate the situation, along with the implications on size and resources.

The Marine Corps is hard pressed to field adequate resources to train and equip its various units. The Corps should shift its nonessential jobs to the Navy, thereby reducing its size, keeping only highly qualified recruits, then providing them with higher quality training and equipment, thereby yielding a more elite organization.

The pool of potential Marines is diminishing.[2] Though the population of the United States continues to grow past 330 million in 2018, the percentage of the population that qualifies as potential recruits is shrinking.

---

[2] See Blake Stilwell, "Here's Why Most Americans Can't Join The Military," *Business Insider*, Last Modified September 28, 2015, http://www.businessinsider.com/heres-why-most-americans-cant-join-the-military-2015-9, Accessed May 6, 2018.

In order to enlist, as of 2019 an individual must:

- Pass an aptitude test, the ASVAB
- Pass a rigorous medical exam
- Be a US citizen or resident alien
- Be between 17 and 29 years old
- Have a High School Diploma
- Meet physical, moral, and mental standards

Each of these requirements reduces the potential eligible population; the medical exam, the Armed Services Vocational Aptitude Battery, and the physical, moral, and mental standards aspects being the most stringent. The minimum ASVAB score is 32 for the Marine Corps, meaning that up to 32% of the population does not qualify for service.[3] The physical standards of the Marine Corps further reduce the potential pool; these standards remain strict both for those trying to qualify and for those who have successfully joined the Marine Corps. Even with looser requirements than the Marine Corps, the US Army is facing a crisis. Asthma, obesity, and physical flaws in particular mean that over two thirds of those of age to join are ineligible.[4] There is no sign that these rates will decrease in the near or distant future. Obesity rates are increasing, and poorer communities are hit the worst. In particular, those in the South and Midwest are overwhelmingly obese. While the Midwest is proportionally represented in the underlying population demographics of the Marine Corps, the South is overrepresented and has historically been one of the

---

[3] Understand ASVAB Score, *Official ASVAB Testing Program Website*, http://official-asvab.com/understand_app.htm, Accessed May 6, 2018.
[4] Natalie Johnson, "Three-Quarters of Young Americans Don't Qualify for Military Service," *The Washington Free Beacon*, Last Modified February 22, 2018, http://freebeacon.com/national-security/three-quarters-young-americans-dont-qualify-military-service/, Accessed May 6, 2018.

most important regions for sourcing both Officers and Recruits.[5]

It should also be noted that we are discussing obesity only, not strength. Potential recruits must be able to pass at least an initial strength test consisting of a 1.5-mile run, pull-ups, and crunches. Not only is obesity a recruiting challenge, it is also a performance challenge because of the correlation between the lack of fitness and injury. Many Marines end up on light duty status or medically separated due to injury. The prevalent obesity and weakness of potential recruits is constantly commented on and is considered a national security issue. For a more extensive synopsis of this problem, I recommend watching Lieutenant General Mark Hertling's 2012 TEDx Talk on obesity – available on YouTube.[6]

The logistical issue is clear with numbers. There are about 320,000 active duty personnel in the Navy and about 99,000 in the reserves. The Air Force has about 320,000 active duty personnel, 100,000 in the Air National Guard, and 69,000 in the Reserves. The Army has over 475,000 active duty personnel, over 340,000 National Guard, and almost 200,000 reservists. The Marine Corps has over 180,000 active duty personnel and some 38,000 reserves. Over 2 million individuals thus enact US foreign policy and defend American interests across the world.

---

[5] Shanea Watkins and James Sherk, "Who Serves in the U.S. Military? The Demographics of Enlisted Troops and Officers," *The Heritage Foundation*, August 21, 2008,
https://www.heritage.org/defense/report/who-serves-the-us-military-the-demographics-Enlisted-troops-and-Officers;
The State of Obesity, "Obesity Rates & Trends," Last Modified June 2018, https://stateofobesity.org/rates/ Accessed May 10, 2018.
[6] As of May 7, 2018, the title on YouTube is Obesity is a National Security Issue: Lieutenant General Mark Hertling at TEDx MidAtlantic 2012

That is certainly a small proportion of the 325 million American citizens, but considering that about 60% of the population is too old to join, only 15% of the military is composed of women, the overall U.S. Labor Force Participation Rate is 63% and Employment Rate is 60% (as of March 2019), and about 75% of individuals don't qualify for military service, the ratio of potential applicants to current personnel increases enormously.[7] In other words, it's not as if 325 million Americans could serve and don't, it's closer to 20 million, and many of these individuals would much rather work as civilians or further their education. If you are smart enough to join and physically healthy, you are more likely to want to go to college than to join the military. Problematically from this perspective, the proportion of individuals who graduate college and join the military is even smaller.

So, the pool of potential applicants, and the quality of those applicants, on average, is decreasing. In order to meet its manpower requirements, the Marine Corps sometimes turns a blind eye to some unqualified Recruits, exemplified by Gardo in my Boot Camp Platoon - a Recruit who could only do two pull-ups by the end of training. Doing so may seem acceptable since these Recruits typically don't have infantry related MOS, and therefore would hypothetically not be relied upon to physically serve the Corps. There are likewise multiple waivers available for applicants, everything

---

[7] U.S. Bureau of Labor Statistics, Civilian Employment-Population Ratio [EMRATIO], retrieved from FRED, Federal Reserve Bank of St. Louis; https://fred.stlouisfed.org/series/EMRATIO, May 6, 2019. U.S. Bureau of Labor Statistics, Civilian Labor Force Participation Rate [CIVPART], retrieved from FRED, Federal Reserve Bank of St. Louis; https://fred.stlouisfed.org/series/CIVPART, May 6, 2019. Kim Parker, Anthony Cilluffo and Renee Stepler, "6 facts about the U.S. military and its changing demographics" *Pew Research Center: Fact Tank, News In Numbers,* Last Modified April 13, 2017, *http*://www.pewresearch.org/fact-tank/2017/04/13/6-facts-about-the-u-s-military-and-its-changing-demographics/ (Accessed September 10, 2018).

from criminal history waivers to medical waivers exists and, though sometimes difficult to obtain, it is possible that many slip through that truly should not.

Theoretically and practically, those not meeting standards are actually a major problem. One of the most famous Marine Corps mottos is "Every Marine a Rifleman." Or, as the 29th Commandant of the Marine Corps, General Gray said, "Every Marine is, first and foremost, a rifleman. All other conditions are secondary." In theory, every Marine should be capable of serving as a basic infantryman. This provides an enormous benefit to the Corps. More critical assets can be deployed without having to protect 'noncombatant' supporting units as much – because Marines should be able to protect themselves - commanders have strategic flexibility in the form of emergency reserves, and because the job of supporting units is to help ground troops accomplish their mission – supporting units are better able to assist those in the action if they have some idea of what they're going through.

In practice, Marines can indeed be required to fulfill the motto. Recently in Afghanistan for example, Marines from various supporting units destroyed an infiltration force of some 15 heavily armed Taliban at Camp Bastion.[8] During major conflicts like Vietnam, Korea, and World War II, supporting units engaged in battle on countless occasions. A famous example was the survival and reinforcement of Fox Company during the Korean War.[9] Without the help of supporting units fighting at the front, that conflict could have turned out

---

[8] Gretel Kovach, "Marines Recall 'Surreal' Attack at Afghan Camp," *The San Diego Tribune*, Last Modified October 6, 2012, http://www.sandiegouniontribune.com/military/sdut-fighting-marines-recount-bastion-attack-2012oct06-story.html (Accessed May 10, 2018).
[9] See Bob Dury and Tom Clavin, *The Last Stand of Fox Company* (New York: Grove Press, 2009).

very differently and the battle of Chosin Reservoir could have been a massacre instead of a grueling victory.

Though most out of standards, weak, or chubby Marines are not directly performing tasks in which they come face to face with the enemy right now, it does not mean that they never will, and it reduces the operational capabilities of a command – especially since a unit is only as strong as its weakest link. As Professor John Kotter writes, "Personnel problems that can be ignored during easy times can cause serious trouble in a tougher, faster-moving, globalizing [world]."[10] Despite the various operations of the Marine Corps in 2017 and 2018, this is a comparative geopolitically "easy time" as the United States is not engaged in a major war, and fewer than 35 service members – across all branches - have died each year for the past half-decade; compare this to the relatively light mortality of recent anti-insurgent wars where several thousand died each year, or a land battle in World War II such as Okinawa where nearly 3,000 Marines died in only a *few months*. Personnel problems could indeed become a serious issue when the Marine Corps is once again called to action on a large scale.

At the same time, not all supporting MOS are actually exposed to direct or indirect combat in any capacity. There is an extremely low probability that Marines deployed in non-hostile territory would even be available in time to serve as strategic reserves for a commander – for example those Marines performing administrative tasks in the United States. Yet these Marines take up valuable spots in the Corps and are provided the same amount of training as their more endangered compatriots. Further, because of the semi-independent nature of the Marine Corps, these Marines

---

[10] John Kotter, *Leading Change* (Cambridge, MA: Harvard Business Press, 2012), pp. 63.

also compete with the other services for budgetary resources. In 2016, the budget of the Marine Corps was about 17% of that of the Navy, and 18% that of the Army, a budget which supported a Marine Corps about half the size of the Navy and one fifth the size of the Army.[11] That year, in spite of its budgetary constraints, the Marine Corps participated in 210 operations, 20 amphibious operations, 160 Theater Security Cooperation events, and 75 exercises across the globe.[12]

As a result of this low budget and high operating tempo, the majority of the Marine Corps is underfunded. I have yet to encounter a unit which has a surplus of cash; the Corps is continually scrounging for funds. Equipment becomes obsolete and too expensive to maintain adequately. Marine light armored vehicles and amphibious landing vehicles are, on average, 41 and 27 years old, respectively as of 2017. Marine aircraft are in worse conditions, 80% of units lack the necessary aircraft for training, and only about 40% of Marine Corps fixed wing and rotary wing aircraft are flyable.[13] Without

---

[11] "Budget of the U.S. Navy and the U.S. Marine Corps from fiscal year 2000 to 2017 (in billion U.S. dollars), *Statista*, https://www.statista.com/statistics/239290/budget-of-the-us-navy-and-the-us-marine-corps/, (Accessed May 10, 2018).
MG Thomas A. Horlander, "Army FY 2017 Budget Overview," *Defense Innovation Marketplace*, February 2016, http://www.defenseinnovationmarketplace.mil/resources/Army%20FY%202017%20Budget%20Overview.pdf, (Accessed May 10, 2018).
[12] Lieutenant General Robert S. Walsh, Brigadier General Joseph Shrader, and John Garner, "Statement on 'Marine Corps Ground Programs' before the Subcommittee on Sea-power," *Committee on Armed Services, U.S. Senate*, June 6, 2017, p. 2, https://www.armed-services.senate.gov/imo/media/doc/Walsh-Shrader-Garner_06-06-17.pdf (accessed May 8, 2018).
[13] "An Assessment of US Military Power: U.S. Marine Corps," Heritage.org, October 5, 2017, https://www.heritage.org/military-strength/assessment-us-military-power/us-marine-corps; General Glenn Walters, "Statement of General Glenn Walters Before the Senate Armed Services Subcommittee on Readiness – On Marine Corps Readiness," February 8, 2017, https://www.armed-

the cash to pay for replacement parts, maintenance personnel duct tape and superglue planes back into service. Pilots receive far less training on their equipment than they should, increasing the probability of mishaps and decreasing their effectiveness. Attempts to sustain vehicles and equipment past their expiration date causes them to malfunction.

The lack of funds is why so many headlines showcasing accident after accident appear in the media, particularly with regards to aircraft.[14] This imperils the crews manning the equipment, the Marines they support, the mission to which they're assigned, and the added variable costs for maintenance, the sum of which could be greater than the costs of the original equipment. That is not even to mention the modernization needed to compete effectively with Russia and China – who unveiled state of the art weapon systems in the past several years that perform either on par or outperform our own capabilities.[15]

---

services.senate.gov/imo/media/doc/Walters_02-08-17.pdf (accessed May 8, 2018).

[14] See for example, Ryan Browne, "US Marine Corps Suffers Third Aviation Incident in Less than 24 Hours," *CNN Politics,* April 4, 2018, https://www.cnn.com/2018/04/04/politics/us-marine-corps-aircraft-incident/index.html (Accessed May 8, 2018);
Sarah Sicard and Jared Keller, "The KC-130 Crash is The Latest Tragedy in the Marines Corps' Worsening Aviation Mishap Crisis," *Task & Purpose*, June 11, 2017, https://taskandpurpose.com/marine-plane-crash-data/ (Accessed May 8, 2018)

[15] Stepan Kravchenko, "Here Are the new Russian Weapons Putin Just Showed Off," *Bloomberg*, March 1, 2018, https://www.bloomberg.com/news/articles/2018-03-01/putin-s-newest-nukes-the-weapons-he-showed-off-in-speech (Accessed May 8, 2018); Jonathan Marcus, "Should Russia's new Armata T-14 Tanks Worry Nato?" *BBC News*, May 30, 2017, http://www.bbc.com/news/world-europe-40083641 (Accessed May 8, 2018); Andrew Roth, "New Russian Stealth Fighter Spotted in Syria," *The Guardian*, February 22, 2018, https://www.theguardian.com/world/2018/feb/22/new-russian-stealth-fighter-spotted-in-syria (Accessed May 8, 2018); Fox Van Allen, "China's Newest Weapons of War," *CBS News*, April 13, 2017,

Increasing spending is not likely to be possible. The amount relative to GDP the United States spends on the military is decreasing in the long run, while spending on entitlements like healthcare and social security has been increasing.[16] An aging and politically active population, along with their rising healthcare costs makes cutting these expenditures difficult, while increasing taxes is a good way to be voted out of office. As a result, the nation takes on debt; the national debt stood at $21.1 trillion as of May 2018 and continues to grow. That debt carries an interest, which in turn has been rising. It is estimated that in a few years, the government will spend more on the interest for its debt than it spends on the military.[17] Without delving into the divisive political arguments engendered by the nation's circumstances, we must understand that military funding is part of a stressed governmental budget and is thereby affected by these conditions. Therefore, to fund Marine Corps operations, a wiser allocation of funds is necessary.

To summarize, the Marine Corps has serious macro personnel and financial supply problems. The Marine Corps cannot direct public or fiscal policy, much less force Americans to 'get in standards.' It can however become more stringent in its standards and simultaneously reduce the accepted number of recruits to remain ahead of the tide. Of course, this would diminish the size of the Marine Corps. In turn, a number of MOS,

---

https://www.cbsnews.com/pictures/chinas-newest-weapons-of-war/ (Accessed May 8, 2018)

[16] Greg Ip, *The Little Book of Economics*, (Hoboken, NJ: John, Wiley and Sons Inc., 2013) pp. 205ff

[17] Nelson D. Schwartz, "As Debt Rises, the Government Will Soon Spend More on Interest Than on the Military," New York Times, September 25, 2018 (Accessed September 26, 2018).
https://www.nytimes.com/2018/09/25/business/economy/us-government-debt-interest.html

which are almost never exposed to conflict could be deleted or shifted to other branches of the military, and their resources reallocated.

The Marine Corps already has a special relationship with the Navy in terms of medical and religious personnel. The Navy provides medical services to the Marine Corps, particularly through Navy corpsmen and providers. These personnel have stood side by side with Marines for over a hundred years. Many are exposed to combat and perform admirably. Navy Chaplains have also served side by side with Marines through many climes and places. Why couldn't the 'garrison' MOS be shunted onto the Navy? Food Service, Distribution Management, Financial Management, Utilities, Logistics, Administrative, and Supply MOS in particular, could be eliminated from the rosters of the Corps. According to the Marine Corps Almanac, the Corps would be up to 15% smaller after displacing solely these jobs. Further cuts can be made by consolidating, merging, and providing cross training between interrelated MOS such as Ground Electronics Maintenance (MOS Code 2800) with Communications (MOS Code 0600), and Organizational Avionics Maintenance (MOS Code 6300) with Intermediate Avionics Maintenance (MOS Code 6400), and Aircraft Maintenance (MOS Code 6000) with its peers Aircraft Maintenance, Rotary Wing (MOS Code 6100) and Aircraft Maintenance, Fixed Wing (MOS Code 6200) would further reduce the total number of required Marines.[18]

Fewer Marines means that productivity per Marine would have to rise. Smaller platoons during Boot Camp and MCT (discussed later in this book) would

---

[18] Headquarters, Marine Corps, *U.S. Marine Corps: Concepts & Programs 2013, America's Expeditionary Force in Readiness* (2013), Chapter 4, particularly pp. 229 and 232.

result in each individual recruit receiving better training. As a consequence, each recruit would be prepared to perform at a level which would increase productivity. Follow-on schools could be more selective and thorough, personnel armed with the ability to think critically would not only be smarter and more productive, they would also be far more flexible and capable of picking up new skills. Thus, personnel in Avionics Maintenance (MOS Code 5900) may learn the overlapping tasks performed by their Communications counterparts. When Marines finish their MOS schools, many end up learning other specialties anyway because of mission demand or missing personnel; for example, 5974s can become acting 5939s and 5979s.

In the words of a RAND Corporation technical report on military productivity, "[cross training and consolidation] increases the utilization of these individuals and reduces the need for additional personnel with more limited skills. These observations suggest that additional training and acquisition of new skills can significantly raise the flexibility of manpower planners and allow the force to perform with fewer personnel."[19] A Marine Corps more dedicated to building and improving junior Marines may even make it more attractive to potential applicants. Rather than joining to become proverbial workers on an assembly line, Junior Enlisted Marines would become thinking, creative, ambitious individuals more prepared for the outside world than ever before.

Such consolidation would make each individual Marine more a generalist and less of a specialist. They would be more easily transferred between tasks of

---

[19] Jennifer Kavanagh, "Determinants of Productivity for Military Personnel, a Review of Findings on the Contribution of Experience, Training, and Aptitude to Military Performance," *RAND Corporation* (Santa Monica, CA: 1981), pp. 25.

different natures and could specialize in the long term as they increase in veterancy and rank. Thus Noncommissioned Officers would become subject matter experts in their specialty, able to train and supervise junior Marines who have skills to understand problems and overcome them using critical thought; an idea somewhat akin to having junior Marines be in an extended corporate 'rotational program', while Noncommissioned Officers rise into specialized subjects after having understood the interrelationship between various tasks. The current system is reversed, Marines begin highly specialized and are forced to become more generalized as they rise in rank and take up managerial roles. Since they practice their MOS less as they rise and conditions of their MOS change – due to new technology or adapted operating procedures – many senior Enlisted have no idea what their Marines are doing. This makes management, supervision, and adaptation difficult in an organization that is top down. Making junior Enlisted more generally useful and senior Enlisted specialized personnel would therefore play directly into the strengths of the military organization and result in better decision making.

The resources previously allocated to these jobs could be redistributed to the cash-starved Corps, while the reduction in size would allow standards to be applied more rigorously, or even increased. The majority of a modern military is not engaged in combat; the majority is at the rear;[20] a substantial displacement of jobs from the Corps would free up considerable space. While the Corps would certainly become more reliant on the Navy, rationalization through centralization is a proven method to reduce runaway bureaucratic costs without diminishing performance.

---

[20] John Keegan, *The Face of Battle* (London: Penguin Books, 1976), pp. 316.

Chapter 2: Strategic Pivot

A more streamlined Marine Corps is compatible with a new strategic initiative given developments across the world. In a sense, the Marine Corps continues to face an existential crisis. The Corps is broadly composed of its own Air Force and Army. Its overlapping capabilities with the Army in particular have made it a target for consolidation or deletion for hundreds of years. Since 1834 the Corps has not been an independent organization; the book *First to Fight* highlights the constant bureaucratic conflicts fought by Marine leadership to keep the Corps in existence. Yet there is a clear solution to the redundancy of the Corps presented by shifting conditions in combat: The Marine Corps can become the nation's premier counterinsurgent fighting force.

In the early 20th century, with the apparent redundancy of the Corps and future conflicts of the U.S. in mind, then Commandant Major General John A. Lejeune pivoted the purpose of the Marine Corps in anticipation of a major Pacific campaign that ended up arising in WWII.[1] He ensured that the Marine Corps would specialize in amphibious landings and the capture of forward operating bases. General Lejeune's foresight proved wise in World War II as Marines spearheaded

---

[1] Robert Heinl, "The U.S. Marine Corps: Author of Modern Amphibious Warfare," in Merrill Bartlett, ed. *Assault from the Sea: Essays on the History of Amphibious Warfare* (Annapolis: Naval Institute Press, 1983), pp. 185-194; General Merill B. Twining, *No Bended Knee: The Battle for Guadalcanal* (New York, Random House Publishing Group, 1996), pp. 2.

attacks against Japanese islands in the island-hopping campaign. The National Security Act of 1947 enshrined the functions of the Marine Corps in accordance with its employment in WWII.

After the war however, the number of applications for major amphibious landings has decreased. In the Korean War, the unorthodox surprise landing at Incheon (1950) was the only major amphibious action of that conflict. In the Vietnam War, the landing at Da Nang (1965) was unopposed, as were the landings in Somalia during Operation Restore Hope (1993). The wars in the Middle East, such as the Gulf War, Iraq, and Afghanistan did not require a complicated beach assault for which the Marine Corps is prepared.

As a result, the post-World War II Corps has been employed in a similar capacity to the Army. Marines are deployed hundreds of miles from the sea, fighting extended campaigns and occupying land for long periods of time. Major campaigns fought by Marine units in modernity include Korea, Vietnam, the Persian Gulf, Iraq, Afghanistan, and Syria. One may therefore ask, what sets the Marine Corps apart from the Army in the 21st century? Further, is it really impossible for the Army to learn how to stage a major amphibious assault? With the resources of the United States and the expertise gathered by the American military over the last century, that answer is a clear no. In fact, historically, the Army has staged more successful amphibious assaults than the Marine Corps including almost all landings in the European theatre in WWII; Army units likewise fought alongside Marines during the island-hopping Pacific campaign. Further, once the beaches have been seized, there is little to distinguish island combat from conventional mainland warfare.[2]

There's a popular misconception that the Marine Corps is the only branch that can be deployed by the President of the United States without Congressional approval. Two pieces of legislature underpin what the President can do – without mentioning Executive Orders issued in times of emergency (where the definition of emergency is open to interpretation). By far the most important is the Constitution. Article II, Section 2 nominates the President as commander in chief of the armed forces. While interpretations of this clause can be a source of conflict, particularly with regards to whether the President should seek Congressional approval for declarations of war, it certainly makes no distinction between the different branches and does not make any mention of Congress. To attempt to exercise some control over the President after the height of the Vietnam War, Congress passed the War Power Resolution in 1973. It requires the president to notify Congress of any troop deployments within two days, and to withdraw any troops after sixty days if Congress has not granted an extension.

Because of the aforementioned conflict with the Constitution, Presidents do not really follow the Resolution. The Constitutionality of this law has not been tested as no organization – Congressional or otherwise – has taken the Executive branch to the Supreme Court over its liberal interpretation of Article II of the Constitution. Note once again, that there is no specific mention of the Marine Corps, the President can deploy forces more or less as he or she dictates; historically those forces have often been Marines since

---

[2] See for instance Sledge's accounts of Peleliu and Okinawa: E. B. Sledge, *With the Old Breed: At Peleliu and Okinawa* (New York, Ballantine Books, 1981).

they are more suitable for rapid deployments, in particular because of their disposition aboard naval vessels and proximity to the coast where they can be quickly conveyed to specific theatres of battle. In the late 20th and early 21st centuries, the Corps sustained its rapid deployment status thanks to the efficiency of the MAGTF, the Marine Air Ground Task Force. The MAGTF combines the ground and air capabilities of the Corps with all required supporting units and assets to quickly deploy and sustain operations almost anywhere in the world, including humanitarian and stabilization missions.

The ability to deploy quickly does not distinguish the Marines enough from the Army. With its vast resources, the Army could theoretically assemble their own MAGTF in concert with the Navy or Air Force, or simply from the hundreds of military bases the army maintains across the world.[3] Further, as I have discussed, the capacity to execute successful amphibious operations has decreased in frequency, and therefore its value has been reduced. The Marine Corps cannot have its raison d'être be based upon a tenuous capability that is maligned with the reality of modern and future warfare. The Marine Corps must pivot its principle mission.

If amphibious assaults can no longer be the distinguishing capability of the Corps, what should be? The Army specializes in conventional ground operations, the Air Force in the air, and the Navy at sea. If three major military branches with far greater resources than the Corps already specialize in each aspect of conventional warfare, how can the Marine Corps fit in? As Clausewitz wrote, "War is the

---

[3] David Vine, "Where in the World is the U.S. Military?" *Politico Magazine*, July 2015, https://www.politico.com/magazine/story/2015/06/us-military-bases-around-the-world-119321, Accessed May 10, 2018.

continuation of politics by other means." The political landscape and potential enemy capabilities must be the determinant of a novel strategy and mission. The department of defense decided in 2018 that, "inter-state strategic competition, not terrorism, is now the primary concern of U.S. national security."[4] This conclusion was based on increased Russian and Chinese assertiveness, coupled with the desire to end the conflicts in the Middle East. But, in spite of the department's shift in focus, inter-state conflict is still at an all-time low, while the number of civil and unconventional conflicts are actually increasing. Inter-state conflicts between Great Powers have almost entirely disappeared.[5]

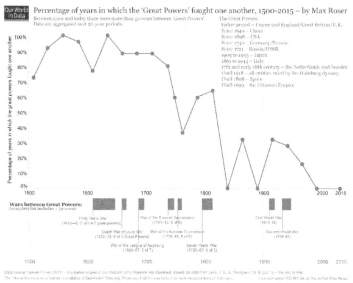

Figure 2: Inter-state conflicts between Great Powers are decreasing in frequency.

[4]James Mattis, "Summary of the 2018 National Defense Strategy of the United States of America: Sharpening the American Military's Competitive Edge,"
https://www.defense.gov/Portals/1/Documents/pubs/2018-National-Defense-Strategy-Summary.pdf, pp. 1.
[5] Max Roser, "War and Peace," *Our World in Data*,
https://ourworldindata.org/war-and-peace (See Chart)

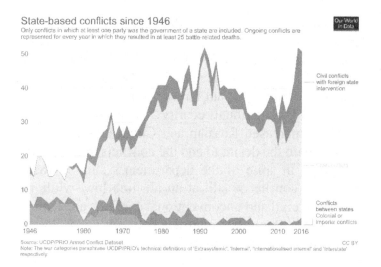

Figure 3: While interstate conflicts decline in frequency, civil conflicts have not.

Overall, the long term trends have been that the world is growing richer, more democratic, more 'rational,' and more peaceful than ever before.[6] In today's world, it is more likely that states will fight proxy wars with one another instead of confronting one another in the open.[7] That goes without mentioning nuclear theory and deterrence, which has also contributed to reducing the number of major interstate wars.[8] Yet while deaths due to conventional wars have almost disappeared, the number of unconventional conflicts and associated deaths from these conflicts are still rampant.

---

[6] John Oneal and Bruce Russett, "The Kantian Peace: The Pacific Benefits of Democracy, Interdependence, and International Organizations 1885-1992," *Source Politics*, Vol. 52, No. 1 (1999), pp. 1ff.

[7] Colin Gray, *Another Bloody Century* (London: Weidenfeld & Nicolson, 2006), pp. 138ff.

[8] Readers interested in these patterns should read Steven Pinker's *The Better Angels of Our Nature* (2011), a work that has highlighted how violence has decreased over both the short and long run.

While interstate conflicts decline in frequency, civil conflicts have not. [9]

Should the present state of military conflicts continue, or proxy wars erupt, the United States is currently heavily reliant on its Special Forces – partly because of the adaptability required in these missions, but mostly because secret missions risk little political blowback and require no congressional oversight. These missions cut through the bureaucratic red tape of military operations. The operations undertaken by American Special Forces include missions to stabilize nations, train and advise friendly forces, counter hostile foreign influence, bolster conventional forces, provide medical and agricultural assistance, leadership decapitation, and strikes at enemy critical vulnerabilities. About 8,000 operatives are active at any time around the world, up 175% since 2001. Commandos are spread across 143 countries, or about 75% of countries in the world.

The propagation of Special Forces is not necessarily a good sign. All the evidence suggests that Special Forces are overstretched across the world. Additionally, every use of a special operations unit is a simultaneous drop in the potential application of that unit somewhere else in the world, which may lead to critical opportunities being bypassed, or to inferior operational capabilities. Every death of an operator is a significant loss to the nation, for they are enormous investments. It costs about $400,000 to train a single SEAL, and over $1.1M a year to maintain a SEAL operative overseas. [10] SEALs and other special operatives take years to train and space is limited – the training pipeline will not easily or quickly replenish lost personnel. Yet many of the

---

[9] ibid.

[10] Thomas Smith, "Money for American Commandos," *Human Events*, April 23, 2008, http://humanevents.com/2008/04/23/money-for-american-commandos/ (Accessed May 11, 2018), adjusted for inflation since 2008.

missions currently undertaken by special operators do not necessarily have to be accomplished by them. While special operations are classified and do not require civilian approval, that comes at the cost of critical thought over strategy, lack of oversight, the potential for policy disasters, and very high monetary expenditures. For those who are deployed, their personal lives become damaged as they are deployed again, and again. The tempo increases the likelihood of mistakes and exposes the best military assets to destruction. More special operatives died in combat in 2016 and 2017 than did conventional soldiers and Marines, even though there are over twenty times more conventional military members than operatives.[11] Policy makers and senior military officials in Washington acknowledge that operations are unsustainable and that commandos do not provide a suitable solution for the problems to which they are often assigned.[12]

This scenario had previously been anticipated in policy papers which warned that "improper employment of SOF (Special Operations Forces) could result in their depletion. SOF require a long lead-time to be effectively fielded. SOF cannot be quickly replaced/reconstituted nor can their capabilities be rapidly expanded. Improper employment of Special Operation resources in purely conventional roles or on inappropriate/inordinately high-risk missions runs the risk of depleting these resources rapidly."[13]

[11] W.J. Hennigan, "The New American Way of War," *Time*, November 30, 2017, http://time.com/5042700/inside-new-american-way-of-war/ (Accessed May 12, 2018)

[12] Vera Bergengruen, "Elite Troops Are Being Worked too Hard and Spread too thin, Military Commander Warns," *McClatchy DC Bureau*, last updated May 5, 2017, http://www.mcclatchydc.com/news/nation-world/national/national-security/article148682644.html, (Accessed May 12, 2018).

[13] T.J. Keating, Doctrine for Joint Special Operations 3-05, December 17, 2003, pp. II – 3; for a more updated version of JP 3-05, see Lieutenant

Low combat deaths for conventional forces are not necessarily a good sign. Training is an incredibly important factor when fighting a war. Commandos are trained extensively and there are never shortages of missions for them to perform. Conventional forces require real combat experience to maintain preparation, to test tactics, and survey equipment. This lesson has not been lost on Russia, which has been testing hardware and practicing tactics in Eastern Ukraine and Syria in recent years. In Syria alone the Russian military has tested over 200 weapons.[14]

A possible solution to the strain felt by commandos is to shunt some of their missions onto the Marine Corps. According to Joint Special Operations policy, the definition of special operations is, "operations conducted in hostile, denied, or politically sensitive environments to achieve military, diplomatic, informational, and/or economic objectives employing military capabilities for which there is no broad conventional force requirement. These operations often require covert, clandestine, or low-visibility capabilities."[15] Yet as was discussed above, many of operations currently undertaken by American commandos are done for political rather than mission specific reasons. Training, assisting, and advising allies, raids, reconnaissance, anti-guerilla, and counterinsurgency operations could be performed by Marines. In addition, policy dictates that Special Forces,

General David Goldfein, Doctrine for Joint Special Operations 3-05, July 16, 2014,
http://www.jcs.mil/Portals/36/Documents/Doctrine/pubs/jp3_05.pdf.
[14] Charles Bybelezer, "How Russia is Using Syria as a Military Guinea Pig," *The Jerusalem Post*, February 28, 2018,
https://www.jpost.com/Middle-East/How-Russia-is-using-Syria-as-a-military-guinea-pig-543839 (Accessed May 12, 2018)
[15] T.J. Keating, Doctrine for Joint Special Operations 3-05, December 17, 2003, 17 Dec. 2003, pp. I-1.

"should not be used for operations whenever conventional forces can accomplish the mission."[16] Central to the successful accomplishment of special operation missions is, "individual and small unit proficiency in a multitude of specialized, often nonconventional combat skills applied with adaptability, improvisation, innovation, and self-reliance,"[17] all of which are traits and skills that should be expected of capable Marines.

With the structure of the MAGTF, the Marine Corps is the best option for reinforcing or replacing beleaguered Special Forces spread across the globe; a Marine Expeditionary Unit, Marine Expeditionary Brigade, or Special Purpose MAGTF would be able to provide all the combat support and logistical capabilities needed to undertake missions across the world at a moment's notice. Further, the Corps has a long and proud history of unconventional and guerilla operations. In a sense, the Corps acted as the original Special Forces for the United States before the rise of modern specialized units in the Second World War.

In the 1800's the Marine Corps fought in the deserts of Libya against an overwhelming enemy, a squad sized element commanded by Lieutenant Presley O'Bannon led an army of mercenaries to capture the city of Derna,[18] an action perhaps similar to that fought by the 5th Special Operations Group in Afghanistan in 2001.[19] The risky operation to protect the diplomatic

---

[16] T.J. Keating, pp. II-2.

[17] T.J. Keating. pp. I-2.

[18] Leo Daugherty III, *The Marine Corps and the State Department: Enduring Partners in United States Foreign Policy, 1798-2007,* (Jefferson, NC: McFarland, 2009), pp. 11ff.

[19] Drew Brooks, "Soldiers Recount True Story Behind '12 Strong,' *US News*, January 27, 2018, https://www.usnews.com/news/best-states/north-carolina/articles/2018-01-27/soldiers-recount-true-story-behind-12-strong (Accessed May 11, 2018).

quarter during the Boxer Rebellion at the turn of the century was undertaken by several hundred Marines. It was there that Dan Daly won his first Medal of Honor when he allegedly killed some 200 Boxers alone in one night.[20] The Corps participated in the major Philippine-American War and insurrection following the annexation of the Philippines in 1898. Above all, it was the Corps that fought the so-called Banana Wars of 1898-1934 in South America and the Caribbean. These conflicts were anything but conventional, and the major lessons extracted were immortalized in the Marine Corps *Small Wars Manual* and revolved around the conduct of guerilla warfare and small unit conflict.[21] It was only with the 20th century and the dramatic increase of Marine Corps numbers during WWI and WWII that the Marine Corps shifted solely to conventional warfare.

The Marine Corps is the logical choice in becoming the nation's preeminent counterinsurgent force. Elite Marines could be rushed across the world, rapidly integrate into their new environments, relieve endangered special operatives, and become the United States' first large-scale hyper adaptive force, ready to deploy across the world at a moment's notice. In a world that sees an increase in insurgencies and unconventional tactics (see Part 1, Chapter 4) it is clear that the United States must develop such a force. It can do so by increasing the number of special forces, or it can repurpose the massive, currently redundant, nonpareil force already at its disposal.

Such a pivot is a major, radical proposition. As a result, the changes I suggest in training are not predicated

---

[20] Katherine Ainsworth, "American Heroes: Sgt. Major Dan Daly, USMC." *U.S. Patriot Tactical*, January 12, 2015, https://blog.uspatriottactical.com/american-heroes-sgt-mjr-dan-daly-usmc/ (Accessed May 11, 2018).
[21] United States Marine Corps, *Small Wars Manual*, 1940.

solely upon this shift but apply equally to conventional war; even if the Marine Corps were to become a primary counterinsurgent force, a major conventional conflict would recall the Corps to fight alongside the other conventional branches. Whether the United States enters a major interstate war, or if counterinsurgency and guerilla conflicts continue to rage on, the Marine Corps is positioned to take the lead on both fronts. Doing so would require a change in training to create Marines that are smarter, faster, and more maneuverable – basically better trained. Thus, even if no strategic pivot is initiated, the changes I suggest would still improve the Marine Corps. To be clear, Basic Training is the fundamental building block to create Marines. A change in training reflects a massive doctrinal shift over what Marines should be like and what Marines should do, which reverberates past training into jobs, deployments, and combat.

Chapter 3: Training Changes Theory

As I discuss in Part II, in Boot Camp we learned to drill, to scream, to follow tight repetitive disciplinary protocols, formal marksmanship, and to perform the simplest of tasks with speed. Other than marksmanship, it is debatable how much the other factors prepare a Marine for modern war. True to the Corps' deep-rooted traditions, these aspects of training came from first-generation warfare in Europe, which was the norm from the 17th to 19th century in Western Europe. Order, control, centralization, and standardization were the hallmarks of first-generation warfare.[1] It is no revelation that trying to directly graft the methods of first-generation warfare to the modern battlefield would result in complete and utter catastrophe.

Yet, there is significant value to each of these elements of training. For instance, there are five purposes to learning close order drill, according to the Marine Corps:

1. Move units from one place to another in a standard, orderly manner.

2. Provide simple formations from which combat formations may be readily assumed.

3. Teach discipline by instilling habits of precision and automatic response to orders.

---

[1] Wlliam S. Lind 'The Theory and Practice of Maneuver Warfare,' in Richard Hooker, ed., *Maneuver Warfare, an Anthology* (Novato, CA: Presidio Press, 1993), pp. 4.

4. Increase the confidence of junior Officers and Noncommissioned Officers through the exercise of command, by the giving of proper commands, and by the control of drilling troops.

5. Give troops an opportunity to handle individual weapons.

While purpose number two is definitely passé, the others still apply. Drill is not easy and requires a tremendous amount of discipline and work to master. The emphasis on screaming in Boot Camp conditions Recruits to perform in the noise Marines may encounter on the battlefield. As one of my Drill Instructors put it, "How the hell is that guy 20 meters away going to know that you need more ammo if he can't hear you?" Discipline is a hallmark of military professionalism and is also one of the principle factors identified by Field Marshall Slim as contributory to success in modern (conventional or unconventional) war.[2] Repeating what to say formally, how to look, and how to act, is what sets Marines apart from civilians and less professional militaries. It is part of the great tradition and culture of the Corps. Performing simple tasks quickly sets a standard for physical movement, increases stamina and trains Marines to respond to orders.

As these are all important hallmarks of training, I do not in any way believe they should be removed. However, three months of training should not be limited to training for first-generation warfare and little else. The impact of this practice is that Recruits become zombies; they have absolutely no common sense. The obsession with first-generation tradition makes Recruits senseless robots, akin to Frederick II's automatons (who were

---

[2] William Slim, *Defeat Into Victory: Battling Japan in Burma and India, 1942-1945* (New York: Cooper Square Press, 2000), pp. 542f

wiped out by Napoleon).[3] I am not the only one to have noticed some of the fallacies in Basic Training. In 2006, Captain Dynan noted similar issues and recommended several tweaks to Boot Camp, none of which seem to have been implemented.[4] I highly recommend that the reader read this source to provide another perspective on this matter beyond the chronicle of my experiences.

In this chapter, I discuss my synopsis of training during Boot Camp. Using resources from the Commandant's reading list, I discuss what the current status of combat is like in the modern world and may be in the future, to then introduce adaptations to basic training. Finally, I examine the feasibility and methods of implementing proposed adaptations to basic training, keeping in mind the limited resources of the Marine Corps.

Purpose of Training
The purpose of recruit training is to develop civilians into basic Marines: to transform Recruits into basic Marines through a thorough indoctrination in the history, customs, and traditions of the Corps, by imbuing them with the mental, moral, and physical foundation necessary for successful service to Corps and Country. The key point lies in the overarching intent: to become basic Marines and receive the necessary training to serve in the Marine Corps effectively. My Senior Drill Instructor once told the platoon that Boot Camp was all about creating a product – the output Marine. As I have argued, the current output Marine is not equipped with the most basic knowledge to operate, much less win, in

---

[3] Martin Van Creveld, *Command in War* (Cambridge, MA: Harvard University Press, 1985), pp. 56.
[4] SP Dynan, "Updating Tradition: Necessary Changes to Marine Corps Recruit Training," *United States Marine Corps, Command and Staff College, Marine Corps Combat Development Command, Marine Corps University*, March 31, 2006.

combat. Though the purpose of *recruit* training is to develop civilians into Marines, the purpose of military training *in general* is to develop forces that can win in combat.[5]

In suggesting changes to basic training, one must ask what is necessary to create the best possible basic Marine, and what are the exigencies of a basic rifleman on the modern battlefield? To a great extent, the infantry is the most important branch of the Corps - every other arm exists to support it. Even in the increasingly mechanized combat of the 20th century, conflict between machines often devolved into struggles among infantry, when well-trained and well-equipped infantry were capable of destroying almost any enemy asset.[6] Currently, units about to deploy go through extensive pre-shipping exercises and their post-basic training does not reflect their actual combat capabilities. This is because of the comparatively leisurely deployments the Corps enjoys today. In major wars of the past, the manpower demands did not allow for as much pre-shipping training, and Boot Camp itself was shortened. During the Vietnam War Boot Camp was 8-9 weeks long instead of today's 12-13, while in World War II it was half that long. It's possible to improve. The Wüttemberg Mountain Battalion in which Erwin Rommel served and with which he achieved some of his most incredible victories in WWI, was formed and trained. in only three months. The high intensity training made it one of the most effective units of the war.[7]

Some of the training I suggest is actually done in the Marine Corps' Sergeants course, prompting one to

---

[5] U.S. Marines, *MCDP1: Warfighting* (North Charleston, SC: Create Space Independent Publishing Platform, 2010), pp. 38.
[6] John Keegan, *The Face of Battle* (London: Penguin Books, 1976), pp. 293.
[7] Erwin Rommel, *Attacks* (Provo, UT: Athena Press, 1979), pp. 91.

ask the necessity of such modifications. A mantra of the Corps is that one must be able to, if necessary, perform the duties of a billet two ranks higher than one's own. The idea is that since leaders can die in battle and leadership vacancies occur often in normal life, subordinates must be capable of filling that role without a reduction in effectiveness. A junior Marine (a Private, Private First Class, or Lance Corporal) is two ranks below a Sergeant and should be able to act as a fire team leader, typically a Corporal, or a squad leader, often a Sergeant. Further, with intelligent and capable subordinates, a team may operate and implement actions more effectively, and make more optimal decisions. Finally, when those junior Marines do get promoted, they will have a more solid baseline from which to lead and improve.

Combat in the Modern World
As discussed in Chapter 2, the Corps faces potential opponents who may employ both conventional and unconventional strategies. For both types of enemies however, similar basic Marines are required. For what should Marines be trained in modern war? I construct my argument by compiling a series of quotations from the Marine Corps' theoretical approach to war, stemming from the Marine Corps Doctrinal Publication (MCDP1) – *Warfighting*, and followed with a practical discussion of both conventional and unconventional warfare at the unit level in modern war in accordance with MCDP1's guiding principles. I aim to show that thoughtful and flexible fire teams and squads should be the bedrock of the Marine Corps.

MCDP1, a text published by the Corps to discuss how the Marine Corps should engage in combat, clearly states that the Marine Corps is dedicated to pursuing maneuver warfare. MCDP1 is anchored in the reality of war and seeks to create a warrior and command culture that thrives in chaos. Chaotic conditions must be harnessed to achieve victory, predicated on decentralization and the development of responsibility and leadership among subordinates.

"We must... be prepared to thrive in an environment of chaos, uncertainty, constant change, and friction. If we can come to terms with those conditions and thereby limit their debilitating effects, we can use them as a weapon against a foe who does not cope as well. In practical terms, this means that we must not strive for certainty before we act, for in doing so we will surrender the initiative and pass up opportunities. We must not try to maintain excessive control over subordinates since this will necessarily slow our tempo and inhibit initiative. We must not attempt to impose precise order on the events of combat since this leads to a formulistic approach to war. We must be prepared to adapt to changing circumstances and exploit opportunities as they arise, rather than adhering insistently to predetermined plans that have outlived their usefulness.... [Warfighting style must be] put into practice during the preparation for war... we cannot expect our subordinates to exercise boldness and initiative in the field when they are accustomed to being over-supervised in the garrison. Whether the mission is training, procuring equipment, administration, or police call, this philosophy should apply."[8]

In the natural chaos of war, decentralized units will be able to seize rapidly fleeting opportunities. The image which comes to my mind is a group of aggressive ants attacking a larger opponent. Each individual ant acts of its own volition, taking advantage of any weakness it sees in the far larger and more powerful opponent.

---

[8] U.S. Marines, *MCDP1: Warfighting,* pp. 53

Though individually stronger, the opponent is unable to maneuver against so much aggression coming from so many angles at all its vulnerabilities and collapses, leaving few, if any, ants as casualties.[9] MCDP1 follows, "a military action is not the monolithic execution of a single decision by a single entity but necessarily involves near-countless independent but interrelated decisions and actions being taken simultaneously throughout the organization. Efforts to fully centralize military operations and to exert complete control by a single decision maker are inconsistent with the intrinsically complex and distributed nature of war." [10]

How can we generate such rapid, quick thinking independent units? With grueling training, forcing leadership to the lowest level, and granting freedom to subordinates to act as they see fit in accordance with their commander's intent. Crucially, the exigencies of modern combat are congruent with the prescriptions of MCDP1.

Conventional Wars
The last major conventional wars fought by the United States were the invasions of Iraq in 1990 and 2003, known as Operations Desert Storm and Iraqi Freedom, respectively. Yet the United States had such an overwhelming advantage in firepower and technology over Iraqi forces that these wars do not provide much foresight into future conventional conflicts. For instance the greatest loss suffered by American forces in the Gulf War was due to a minor technical malfunction in an anti-ballistic missile system, which resulted in the destruction

---

[9] For an entertaining example of this scenario, see Monster Bug Wars – Official Channel, *Army Ants vs. Rainforest Land Crab | MONSTER BUG WARS,* Oct. 14, 2015,
https://www.youtube.com/watch?v=9JniO9aQmLY&t=3s, Accessed May 23, 2018.
[10] U.S. Marines, *MCDP1: Warfighting,* pp. 10f

of a temporary barracks and the deaths of its occupants.[11] Thus the last 'difficult' conventional wars the U.S. participated in were the Korean and Second World Wars, to which we will return shortly.

Since then, the United States has been overwhelmingly reliant on its greater firepower, notably the capabilities of the air force. Other than the quagmire of Vietnam, which could hardly be considered a conventional war, airpower was the core of American marginal advantage. Airpower won the Iraq, Gulf, and Kosovo Wars,[12] airstrikes constituted the basis of American interventions in the recent or current conflicts of Libya, Syria, Iraq, and Afghanistan. It's clear that the U.S. has, for about a half century, relied on superior airpower to fight its wars. I distinctly recall at MCT being presented with various theoretical scenarios by combat instructors. I was surprised that when asked what they would do about any hostile activity, all instructors said they'd more or less hunker down and call artillery or air strikes upon the enemy.

If history teaches us one thing, it's that marginal advantage is fickle, it slips away. Adversaries adapt their assets and strategies to negate hostile advantages. Both Athens and Sparta provide particularly salient examples. Athens, the great sea power, was defeated at sea, culminating in their calamitous loss at Aegospotami in 405 BC. In turn Sparta was the land power par excellence of Ancient Greece. They were crushed on the plains of Leuctra in 371 BC. Germany likewise did not or was unable to deploy the tank in WW1; by WW2 Germany had built significant numbers and employed them in a

---

[11] Richard S. Lowry, *The Gulf War Chronicles: A Military History of the First War with Iraq* (Lincoln, NE: iUniverse Inc., 2008), pp. 124.

[12] Robert Kaplan, *The Revenge of Geography: What the Map Tells us About Coming Conflicts and the Battle Against Fate* (New York: Random House, 2012), pp. 16.

novel way to destroy their neighbors, in spite of the fact that France alone had more men, tanks, and artillery (and of better quality) than Germany![13] Technology employed in war, in particular, shifts with great rapidity. For any nation engaged in a conflict, it must abide by the red queen hypothesis to maintain any superiority at all: militaries must constantly evolve and improve to just to maintain an advantage and survive.[14] The technological advantage the U.S. relies upon is particularly vulnerable given the rate of technological improvement and change in the modern era. Research suggests that military and civilian patents diffuse at the same rate, so progress in the civilian sector can serve as a general benchmark for technological progress.[15] While less than one million patents were filed worldwide in 1985, over 3 million were filed in 2017.[16] Transistors also provide a proxy to measure technological change, which has followed Moore's law quite closely.

[13] Thomas Rodney Christofferson and Michael Scott Christofferson, *France During World War II: From Defeat to Liberation* (New York: Fordham University Press, 2006), pp. 18.
[14] Leigh Van Valen, "A New Evolutionary Law," *Evolutionary Theory* Vol. 1, pp. 16; see James McDonough, 'The Operational Art, Quo Vadis?' in Richard Hooker, *Maneuver Warfare, an Analogy* (Novato, CA: Presidio Press, 1993), pp. 114.
[15] Jon Schmid, "The Diffusion of Military Technology, Defense and Peace Economics," *Doi,* 2017, pp. 12.
[16] I invite the curious reader to look into the accelerating pace of technological change by perusing the statistics made available by the World Intellectual Property Organization (WIPO)

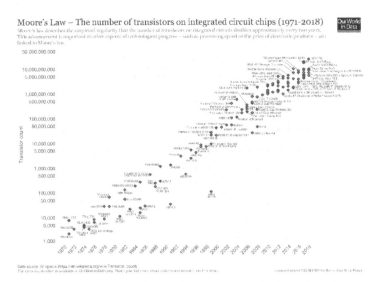

Figure 4: Capacity for transistors as a proxy for technological progress; capacity tends to double every two years.

More troubling geopolitically for the U.S, China received more patent applications than the combined number received by the U.S., Europe, Korea, and Japan in 2018. It should therefore not be surprising that both China and Russia have narrowed or overcome the military technological gap that once existed between the U.S. and its potential adversaries. The total tonnage of new warships China put to sea between 2012 and 2018 is greater than the sum tonnage of the entire French navy![17] Further, developments in small weapons and

---

https://www.independent.co.uk/news/world/politics/china-russia-us-military-challenge-western-allies-nato-strategy-war-military-balance-a8209771.html, Accessed May 25, 2018.

[17] Martin Van Creveld, *Command in War* (Cambridge, MA: Harvard University Press, 1985), pp. 205.

[17] Though we do not discuss China in depth in this book, with its huge economy and massive manpower, China could be the greatest conventional military threat to the United States. Yet it is not discussed

electronic warfare capabilities could even the playing field, just as developments in missile technologies diminished the marginal advantage of tanks in the last 20th century, exemplified by the salvo of Egyptian anti-tank missiles that wiped out two-thirds of General Mandler's armored division in 1973.[18] In future conventional wars, it is therefore unlikely that the U.S. will have the same crushing technological advantage enjoyed in the wake of WWII.

We must draw on as many possible sources of information to forecast the possible requirements of a conventional conflict: war games and research, the Syrian Civil War (2011-Present), and the Donbass War (2014-Present).[19] The most famous American war game in recent history – the Millennium Challenge 2002 - was captured in Malcolm Gladwell's *Blink: The Power of Thinking Without Thinking*. There were two teams, blue team and red team, the blue team represented the United States while the red team represented a rogue military

---

because it is not diametrically opposed to the US – rather it seeks to carve out its own orbit in the world. By all measures, China seeks to assert a sphere of influence in the South China Sea and gain influence across Eurasia through the "Belt and Road" Initiative. For readers interested in this topic, I recommend Robert Kaplan's *The Revenge of Geography.* [17] Kim Sengupta, "China and Russia are catching up with military power of US and West, say leading defense experts," *Independent,* February 14, 2018, https://www.independent.co.uk/news/world/politics/china-russia-us-military-challenge-western-allies-nato-strategy-war-military-balance-a8209771.html, Accessed May 25, 2018.

[18] Martin Van Creveld, *Command in War* (Cambridge, MA: Harvard University Press, 1985), pp. 205.

[19] Though we do not discuss China in depth in this book, with its huge economy and massive manpower, China could be the greatest conventional military threat to the United States. Yet it is not discussed because it is not diametrically opposed to the US – rather it seeks to carve out its own orbit in the world. By all measures, China seeks to assert a sphere of influence in the South China Sea and gain influence across Eurasia through the "Belt and Road" Initiative. For readers interested in this topic, I recommend Robert Kaplan's *The Revenge of Geography.*

commander with strong religious and ethnic support, who sponsored terrorist organizations. Needless to say, blue team had overwhelming resources, information, intelligence, and capabilities: everything could be analyzed in depth. In Gladwell's words, "they had every toy in the Pentagon's arsenal."[20] In contrast, red team was highly limited, with capabilities similar to Iran or Iraq in the early 2000's. Red team employed a series of unconventional tactics to evade the power of blue team, then struck in unorthodox and rapid attacks that crippled and destroyed enormous amounts of blue team's assets. It was a catastrophe for the blues, a stunning success for the reds.[21] The tools that allowed blue team to know and see all had failed. Interestingly, red team employed a command and control philosophy similar to that espoused by MCDP1 (while blue team did not): commanders provided guidance and intent and left decision-making to their subordinates. Ground operators were to use their own initiative and insight to make decisions that resulted in extremely rapid cognition and a high tempo. [22]

What's perhaps most surprising is that unconventional strategies win wars with increasing frequency, and that weaker actors are in fact not overwhelmed by their powerful conventional foes. Professor Ivan Arreguín-Toft describes this trend by analyzing conflicts over two centuries. The null hypothesis was "if power implies victory in war, then weak actors should almost never win against stronger opponents, especially when the gap in relative power is very large." Yet, as he shows, history does not prove this hypothesis correct. Arreguín-Toft surveyed 197 conflicts

---

[20] Malcolm Gladwell, *Blink: The Power of Thinking Without Thinking* (New York: Back Bay Books, 2007) pp. 49.
[21] Malcolm Gladwell, pp. 51.
[22] Malcolm Gladwell, pp. 55.

fought between lopsided opponents – where one opponent has more than 5 to 1 material advantages. He found that over time, weaker actors were winning wars at a greater rate. Other research had suggested this was caused by political weakness at home and that the actor with the most resolve would win the war, without regard to material resources. Thus, though they have the ability to fight it out and conquer, powerful nations lose small wars because their constituencies force a withdrawal before military victory, exemplified by the American public in the Vietnam War.[23] Arreguín-Toft's research suggested that the cause of defeat was not an unreliable constituency but the strategies employed by the belligerents, if the strong actor uses a conventional approach while their weak opponents use an unconventional approach (like guerilla warfare), the weak actor has 2 to 1 odds of winning. In line with this observation, the number of terrorist attacks and related death tolls have increased in the past decade – particularly after 9/11.[24] In what is perhaps a social Darwinian phenomena of war, the ratio at which weak opponents fight using unconventional to conventional approaches has increased dramatically in modern times.[25] In other words, weak opponents are smart, adaptable, and ever more capable of winning conflicts. The policy implication is clearly that the U.S. must be prepared to fight asymmetrical wars as much as conventional wars. I realize this discussion of war games and theory has looped from conventional war into

[23] Andrew J.R. Mack, "Why Big Nations Lose Small Wars: The Politics of Asymmetric Conflict," *World Politics*, Vol. 27, No. 2 (1975), pp. 175–200.
[24] See, Megan Smith and Sean Zeigler, "Terrorism before and after 9/11 – a more dangerous world?" *Research and Politics,* Vol. 4, No. 4 (2017).
[25] Ivan Arreguín-Toft, "How the Weak Win Wars: A Theory of Asymmetric Conflict," *International Security*, Vol. 26, No. 1 (2001), pp. 93–128.

unconventional war, when my goal was to discuss conventional wars. Yet this discussion also suggests that unconventional strategies, particularly fast-striking small unit tactics, win wars against both strong conventional and weak opponents. As of 2018, the United States spends almost 9 times what Russia spends on its military, and 3 times as much as China. In fact, the U.S. spends almost as much as the sum of the next 15 other countries, and that is still only a relatively small fraction of its GDP[26]. Because of its economy, the United States should currently be able to sustain enormous material losses compared to its potential adversaries (though I will return to this point later), suggesting that these adversaries may have to fight in a fully or semi unconventional manner, even though they too are states – a practice very well known to Russia, as exemplified by its combined conventional and guerilla strategies during Napoleon's 1812 invasion and the Great Patriotic War.

In Russia's most recent proxy wars – in Syria and Ukraine – we note the advancement of its conventional capabilities and its use of modern technologies to improve accuracy and efficiency of those capabilities. In Syria, Russia has employed a rather similar strategy as the U.S., deploying Special Forces, mercenaries, and airpower to support their Syrian allies. Yet beyond showcasing their new weapon systems, Russia has not done anything truly tactically surprising in this theatre. A recent battle between Russian mercenaries and American Special Forces showed that American air

---

[26] As measured by dollar value. If measured by purchasing power parity, Russia's defense budget nearly triples while China's almost doubles. From this perspective, the relative material strength of the United States' rivals increases tremendously, dramatically increasing the need for better trained, smarter personnel.

power still reigned supreme in this theatre.[27] The most innovative weapon and tactic has been the use of cyber warfare to inhibit or destroy American electronic capabilities, also employed in the Donbass War.[28] It should be noted that if the U.S. is unable to surmount this restriction on its electronic systems (like communications and radar), it would find itself in a similar position to that of red team in the Millennial War Games, the U.S. would have to adapt around it. Losing these capabilities would be a tactical loss, but not a checkmate.

The Donbass War and related training provide a better view of the present nature of conventional conflicts because Russia was involved more directly. The most notable use of Russian forces seems to be using 'Shock Fire' tactics, and semi-autonomous battalion tactical groups. By and large however, it appears that the tactics are still quite similar to those of the mid-20th century.[29] Shock Fire tactics appear to consist of the integration of UAV's (unmanned aerial vehicles – often called drones), radars, command and control capabilities, personnel, tanks, and artillery to locate and destroy targets. An example of this tactic in exercise took place in the winter of 2016; tanks emerged from their

---

[27] Thomas Gibbons-Neff, "How a Four Hour Battle Between Russian Mercenaries and U.S. Special Commandos Unfolded in Syria," The *New York Times*, May 24, 2018, https://www.nytimes.com/2018/05/24/world/middleeast/american-commandos-russian-mercenaries-syria.html, Accessed May 28, 2018.
[28] Anna Varfolomeeva, "Signaling Strength: Russia's Real Syria Strength is Electronic Warfare Against the US," *The Defense Post,* May 1, 2018, https://thedefensepost.com/2018/05/01/russia-syria-electronic-warfare/ Accessed May 28, 2018; David Brennan, "Russia is Attacking U.S. Forces with Electronic Weapons in Syria Every Day, General Says," *Newsweek,* March 25, 2018 http://www.newsweek.com/russia-attacking-us-forces-electronic-weapons-syria-daily-general-says-900461, Accessed May 29, 2018.
[29] See Gordon Rottman and Peter Dennis, *World War II Infantry Assault Tactics* (Oxford: Osprey Publishing, 2008).

concealed positions to fire, then quickly moved away again as the hypothesized enemy artillery targeted their last positions. Once the artillery fired, Russian UAV's and radars located and transmitted their location to their own artillery to neutralize the threat. On one hand, this isn't a particular mind-blowing tactic – the U.S. also employs drones, while the Marine Corps plans to equip every infantry squad with a quadcopter.[30] They also employ advanced technologies to connect battlefield assets and destroy enemy targets. [31] It does however once again suggest that Russia has narrowed or eliminated much of the operational gap between itself and the U.S.

So-called battalion tactical groups have underpinned Russia's attacks in the Ukraine. They are semi-independent units that flank and encircle an enemy location, then destroy it through attrition in a siege. Russian forces used cyber warfare capabilities to knock out Ukrainian equipment, and artillery fire based on drone data to wipe out targets.[32] It was a slow but remarkably effective tactic when used against the Ukrainians in Zelenopillya, Ilovaisk, the Second battle of Donetsk Airport, and Debal'tseve, all in 2014 and 2015.

Though indirect artillery accuracy has improved, and electronic assets are integrated more effectively in Russian armed forces, the tactics and general capabilities would not have been foreign to Western generals in World War II. Indeed, there are remarkably few

---

[30] Kyle Mizokami, "The Marine Corps' Latest Weapon is a Quadcopter," *Popular Mechanics,* February 9, 2018, https://www.popularmechanics.com/military/a16762519/the-marine-corps-latest-weapon-is-a-quadcopter/, Accessed May 29, 2018.
[31] Roger McDermott, "Russia's Armed Forces Rehearse New 'Shock-Fire' Tactics," *Eurasia Daily Monitor* Vol. 15, Issue 34 (2016).
[32] Amos Fox, "The Russian-Ukrainian War: Understanding the Dust Clouds on the Battlefield," *Modern War Institute*, January 17, 2017, https://mwi.usma.edu/russian-ukrainian-war-understanding-dust-clouds-battlefield/, Accessed May 28, 2018.

differences between the war-fighting methods of today and World War II. As Van Creveld writes, "Taking 1944-5 as our starting point, the most important distinctions continued to be among warfare as waged on land, in the air, and at sea. On land, the most important pieces of equipment continued to be the tank, the armored personnel carrier, and artillery. All of which of course depended on the mechanized transport of supplies... At sea, the most important type of ship continued to be the aircraft carrier. Aircraft were essentially larger, faster, and more powerful versions of their predecessors.

Many entirely new weapons, from attack helicopters to cruise missiles to various kinds of surface-to-surface, air-to-surface, surface-to-air, and air-to-air missiles were also used, but whether they really transformed operations in the same way, and to the same extent, as the internal combustion engine did between 1919 and 1939 is debatable. For what it is worth, I once asked a roomful of military experts... whether they could name any essential differences between the operations of General Norman Schwarzkopf in Iraq in 1991 and those of George S. Patton in France in 1944-45. They could not." [33]

It is worth noting that conventional wars of attrition like World War II are won by the opponent with the most resources; the one who can absorb the most damage and deal the most havoc upon the enemy, which is the greatest reason the allies won both WWI and WWII.[34] Russia's population is relatively stable but peaked in 1991 and, though Russia is larger than the

---

[33] Martin Van Creveld, *The Changing Face of War*, pp. 198f.
[34] Stephen Broadberry and Mark Harrison, *The Economics of World War I* (Cambridge: Cambridge University Press, 2005), pp. 3-40; Mark Harrison, *The Economics of World War II: Six Great Powers in International Comparison* (Cambridge: Cambridge University Press, 1998), pp. 1-40.

U.S., it has less than half of its population (while the population of the U.S. continues to increase). Russian attempts to destroy American armies would likely result in their defeat as long as American logistical capabilities are unharmed (i.e. do not suffer Charles XII's fate in 1708-9, Napoleon's in 1812, or Hitler's in 1943).[35]

The war-fighting similarities between modern day and mid-twentieth century war brings us inevitably to discuss the most effective techniques of the early 20th century.[36] On the surface, the tactics of the First and Second World Wars were entirely determined by the underlying technologies of airpower, artillery, automatic weapons, and armored vehicles. Delving deeper however, those technologies did not necessarily correlate with the best tactics of the war. The allies won not because they were better at fighting but because of the sheer comparative enormity of their resources.[37] And yet, though Germany lost the war, it won the tactical battle using tenants of what we now call maneuver warfare; German forces were by far tactically the most proficient of any power, they were superbly trained and brilliantly adaptive.[38]

Examples of maneuver warfare include Rommel's operations in WWI, Brusilov's offensive in 1916, the use of Stoßtruppen and the beginning of operation Michael in 1918, the Blitzkrieg in 1939-40, operation Compass in 1941 and Israeli military operations in the late 20th century. At its core, the philosophy of maneuver warfare simply states aphorisms

---

[35] See Robert Kaplan's macro evaluation of Russia and its vulnerabilities in *The Revenge of Geography: What the Map Tells us About Coming Conflicts and the Battle Against Fate* (New York: Random House Publishing, 2012), pp. 175ff.
[36] Colin Gray, *Another Bloody Century* (London: Weidenfeld & Nicolson, 2006), pp. 189.
[37] Martin Van Creveld, *The Changing Face of War*, pp. 161ff
[38] Martin Van Creveld, pp. 157ff

of war and suggests that a capable and thoughtful unit can take advantage of those principles. As such, the bases of maneuver warfare are to be found in nearly every major military victory in the past – save perhaps for some first generational wars in the 18th and 19th century.

Maneuver warfare involves a fast tempo to maintain the initiative and exploit enemy weaknesses, destroying vital assets while bypassing enemy strengths. In other words, don't let the enemy dictate your own actions, maneuver intelligently so that you use your assets in the best way possible to achieve victory; maneuver warfare contrasts with attrition-based warfare, which involves destroying or capturing enemy assets, terrain, and strong points, exemplified by British and French operations during WWI. To contrast the two, consider an important castle in the Middle Ages. An "attritionist" would likely besiege it, while a "maneuverist" would bypass the strongpoint, attacking vulnerable assets like convoys and fields, or just continue past the castle unless it was absolutely necessary to attack; even then subterfuge and ruses would be preferable to a direct assault.

The simple tenets of maneuver warfare can be found in most great tactics and strategy over time, even in the ancient world. Alexander the Great, Agesilaus II, Hannibal, Scipio Africanus, Epaminondas, Sulla, Cyrus the Great, Philip II, and Caesar are just a few of the greatest generals of the ancient world who used terrain, thoughtful maneuvering, and brilliant tactics to mitigate the strengths of opponents to create and exploit vulnerabilities. However, the purpose of this chapter is not to discuss the nature or benefits of maneuver warfare,[39] which are generally accepted by the Corps to

---

[39] For this, I would point a curious reader to Richard Hooker, ed., *Maneuver Warfare, an Anthology* (Novato, CA: Presidio Press, 1993);

be superior and more effective than attrition-based warfare, but to discuss the prerequisites necessary to employ it.

Maneuver warfare is based upon decentralization, a factor that coincides with the increasing size of the modern battlefield. Armies have grown enormously over the past several hundred years, while the technologies used have increased the distance over which war is fought – to the point of including every asset of an enemy including his industry, population, and natural resources; a point noted by Keegan in 1976, "the dimensions of modern battles, perhaps 100 miles wide by twenty deep, perhaps even more, [are] certainly thirty to fifty times as large as those of the eighteenth century..."[40] Modern battles fought in the 21st century are likely to be even larger, not counting the titanic theaters of operations that could very well grow into space and virtual dimensions due to cyberwarfare.[41]

The moment an army becomes large and dispersed is the moment that over-centralization becomes an extreme liability. Fronts become too large and there are too many assets to be controlled by any one individual. "Even when the terrain was perfectly flat and open, it was no longer always possible for a commander to watch his entire front without standing so far back as to make effective command impractical."[42]

A battlefield is not a game of chess – all units can be moving at all times, and one cannot see the enemy's movements; one's own units can move on their own, and

U.S. Marines, *MCDP1: Warfighting* (North Charleston, SC: Create Space Independent Publishing Platform, 2010).
[40] John Keegan, *The Face of Battle* (London: Penguin Books, 1976), pp. 315.
[41] Thomas Hammes, *The Sling and the Stone: on War in the 21st Century* (St Paul, MN: Zenith Press, 2006), pp.2.
[42] Martin Van Creveld, *Command in War* (Cambridge, MA: Harvard University Press, 1985), pp. 53.

communication with them is difficult, slow, or impossible; adversaries do not begin equal, and rules are neither preset nor fixed. Predictions often mean little. In the chess analogy, one may be tempted to extrapolate the movement of the entire enemy force based upon the movement of a pawn. Major victories in the past rarely arose from a brilliant plan by a general, but from the actions of units that seized an opportunity; a bishop saw an opening and captured the King. As Von Schell wrote, "A lack of knowledge of the enemy's intentions and actions may always be expected in open warfare. In open warfare, we will never know exactly where the enemy is, how strong he is, or what he intends to do. War is not as easy as a map problem. Leaders must nearly always issue orders without exact information. Our mission and our will are often the only things untouched by obscurity. These will frequently form our only basis for an order. If a leader awaits complete information before issuing an order, he will never issue one." [43]

Even the geniuses on the orchestrated battlefield of first-generation warfare – Frederick II and Napoleon – won their greatest battles by accident, independent of the 'genius' of their leaders. Frederick's cavalry acted on its own and was overwhelmingly responsible for achieving victory while the Prussian infantry struggled to drill correctly. [44] Studying Napoleon's behavior and decisions in battle, it's quite clear he had very little control over the actions of his men and won major battles such as Jena in 1806 without having played a beneficial role whatsoever! There, he knew nothing of the main action during the day, had forgotten two of his corps, issued little to no orders to two others, and had no idea

---

[43] Adolf Von Schell, *Battle Leadership*, (Battleboro, VT: Echo Point Books, 1933), pp. 31.
[44] Van Creveld, pp. 55.

what the fifth corps was doing. So, what is the proper response if no control can effectively be established?

Van Creveld specifically discusses this subject as he sought the most effective form of battle leadership over several centuries. "the only [solution to the command question] that consistently produced victory over a period of centuries and almost regardless of the commander's personality, was the Roman one: that is, a command system, not based on any real technical superiority, that relied on standardized formations, proper organization at the lowest level, a fixed repertoire of tactical movements, and the diffusion of authority throughout the army in order to greatly reduce the need for detailed control." [45]

For "Roman legions in battle scarcely needed a commander in order to gain victory; time and again (Zama, Cynoscephalae, Thermopylae, Magnesia, and Caesar's battle against the Belgians are good examples) the sources mention centurions, or else military tribunes, field-grade Officers all, who "knew what to do" and "judging on the spur of the moment" came to their comrades' aid, or closed a legion's shattered ranks, or took a number of maniples and, apparently acting on their own initiative, carried out an outflanking movement." [46]

Those with the most information on current circumstances are those closest to the action. These commands are more in tune with conditions and can take advantage of present opportunities.[47] The greatest potential drawback is losing sight of the greater picture; but in a modern Marine Corps five paragraph order, that greater picture is provided by commander's intent. This point was understood by the Prussian army as it recast

[45] Van Creveld, pp. 56.
[46] Van Creveld, pp. 46.
[47] Erwin Rommel, *Attacks*, Provo, UT: Athena Press, 1979), pp. 24.

itself after Napoleon's crushing victories in what was called *Auftragstaktik* – the basis of what we now call maneuver warfare. In the Prussian army, "the importance of the mission was reduced. The importance of the commander's intent was very much more emphasized… When unexpected situations sprang up, commanders could more easily act on their own – as long as they stayed within their commander's intent."[48] Of course this would only be possible if subordinates were trained to a high degree of competency.[49] Today, not only are battlefields growing in size – the dispersion between units is also increasing, leading to even less possible centralized control.[50] In other words, as the battlefield continues to increase in scale and scope, centralization is ever more impossible.[51] Subordinates must be intelligent

---

[48] Franz Uhle-Wettler, '*Auftragstaktik:* Mission Orders and the German Experience,' in Richard Hooker, ed., *Maneuver Warfare, an Anthology* (Novato, CA: Presidio Press, 1993), pp. 242.

[49] Franz Uhle-Wettler, pp. 243.

[50] leaders at all levels [must] demonstrate sound and timely judgment. Initiative becomes an essential condition of competence among commanders." U.S. Marines, *MCDP1: Warfighting* (North Charleston, SC: Create Space Independent Publishing Platform, 2010), pp. 53.

[51] Van Creveld completes his work with the following conclusion on effective battlefield commands: "Attempting to generalize from the historical experience studied here, I suggest that there are five such implication, all interacting with each other: (a) the need for decision thresholds to be fixed as far down the hierarchy as possible, and for freedom of action at the bottom of the military structure; (b) the need for an organization that will make such low-"decision thresholds possible by providing self-contained units at a fairly low level; (c) the need for a regular reporting and information-transmission system working both from the top down and from the bottom- up; (d) the need for the active search of information by headquarters in order to supplement the information routinely sent to it by the units at its command; and (e) the need to maintain an informal, as well as a formal, network of communications inside the organization. The fact that, historically speaking, those armies have been most successful which did not turn their troops into automatons, did not attempt to control everything from the top, and allowed subordinate commanders considerable latitude has been abundantly demonstrated… The realization that certainty is the product of time as well as of information, and the consequent to do with less of the latter in

and capable, to have the same critical eye of an enemy, timing, and terrain that heroic warrior generals of the ancient, medieval, and renaissance such as Gustavus Adolphus or Philip II had in a melee.

To successfully implement *Auftragstaktik* the underlying organization has to be flexible and open. Testimonies of foreign observers after the brilliant German unification wars of the mid-19th century and Franco-Prussian War clearly show the basis of this performance:

"Nowhere in this world is independence of thought and freedom of decision as much groomed and supported as in the German army, from the corps commanders down to the last NCO. A Russian general who had been an observer of the Franco-German war arrived, in his two-volume account of the war, at the following conclusion: 'At the root of the German victory is an unbelievable readiness to act independently, a readiness displayed at all levels down to the very lowest and displayed on the battlefield as well as in other matters. Finally, shortly after the Franco-German war, a French lecturer told the students of the École Supérieure de Guerre, all of whom must have participated in the war: Common among the (German) Officers was the firm resolve to retain the initiative by all means… NCOs and soldiers were exhorted, even obliged to think independently, to examine matters and to form their own opinions. These NCOs were the backbone of the Prussian army… their special role, supported by a

---

order to save the former; the postulation by higher headquarters of minimum, rather than maximum, objectives; the freedom granted junior commanders to select their own way to the objective in accordance with the situation on the spot, thus cutting down on the amount of data processing required; and the willingness of superior headquarters to refrain from ordering about their subordinates' subordinates… are indispensable…" Van Creveld, pp. 270.

respect for them unknown in other armies, secured them an honorable and envied position."[52]

The reason command has 'gone down the ranks' despite advances in communication technology (i.e. telegraphy, telephone, radio, satellite), is that weapons simultaneously increased in range and became deadlier. Packed formations to protect against cavalry were not only no longer needed, but became liabilities due to modern rifles, automatic weapons, and artillery. As Van Creveld writes, "breech-loading made firing from a crouching or prone position practicable; thus, troops were able to look for cover. In the absence of novel means of tactical communication, the spreading out of troops and the altered firing position led to a sharp decrease in the amount of control that junior commanders were able to exercise over their men... Entire armies turned into clouds of uncontrollable skirmishers, especially when on the attack."[53] Of course the adoption of the Kammerlader, Dreyse needle gun, Chassepot, and other breech loading rifles of the 19th century occurred simultaneously with developments in other weapon systems, notably artillery, such as the Armstrong gun and French '75, and automatic weapons like the *Mitrailleuse*, Gatling gun, and Maxim gun.[54]

As all the armies of the late 19th and early 20th century found out, dense formations led to catastrophes. The innumerable German columns attacking Belgium and northern France were mowed down by small arms and artillery fire; British infantry units were wiped out as they walked across the valley of the Somme; a third of a million Frenchmen died in the first month of the war as

---

[52] Franz Uhle-Wettler, pp. 241.

[53] Martin Van Creveld, *Command in War* (Cambridge, MA: Harvard University Press, 1985), pp. 107.

[54] Max Boot, *War Made New: Weapons, Warriors, and the Making of the Modern World* (London: Penguin Books, 2006), pp. 149ff.

they attempted to implement the philosophy of *attaque à outrance*; company after company of poorly supplied Russian infantry were obliterated during the invasion of East Prussia.[55] Even when basic dispersion was implemented, artillery still destroyed swathes of infantry as Rommel noted, "with the increased power of modern weapons, increased dispersion and digging of foxholes is vital to the safety of any unit."[56] Indeed the majority of deaths in WWI, WWII, and Korea were caused by artillery.[57] Such a realization was already accepted in World War I, which proved that firepower and maneuver, not numbers, won battles;[58] a lesson that had to be learned and relearned time and again by militaries across the world.

Now of course, weapons have become ever more precise and deadly.[59] While artillery development in the post WWII world had been rather neglected, it has clearly regained its place on the battlefield.[60] Missiles, small arms, automatic weapons, and armored vehicles are far superior to their 20th century predecessors. Artillery can move and deploy extremely quickly and destroy precise targets, at enormous ranges far beyond their line of sight. Guided missiles, such as the Tomahawk cruise missile, don't even require the presence of nearby installations to destroy enemy targets,

---

[55] See Barbara Tuchman, *The Guns of August: The Outbreak of World War I* (New York: Random House, 2009) for a summary of the initial massed infantry tactics colliding with modern weapon systems in WWI.
[56] Erwin Rommel, *Attacks*, Provo, UT: Athena Press, 1979), pp. 24.
[57] Christopher Bellamy, *Red God of War: Soviet Artillery and Rocket Forces* (London, 1986), pp. 1ff.
[58] Timothy Lupfer, "The Dynamics of Doctrine: The Change in German Tactical Doctrine During the First World War," Combat Studies Institute, U.S. Army Command and General Staff College, 1981.
[59] Max Boot, *War Made New: Weapons, Warriors, and the Making of the Modern World* (London: Penguin Books, 2006), pp. 318ff.
[60] J.B.A. Bailey, *Field Artillery and Fire Power* (Oxford, UK: Military Press Ltd, 2009), Chapter I.

and neither do long-range aircraft like TU-160's and UAV's such as the infamous Reaper drone. As anti-Western factions have found across the world, any formation in the open is vulnerable, and both hard targets and armored columns can be rapidly eradicated.[61]

If we focus momentarily on the Marine Corps' specialty – amphibious assaults – we note the immense casualties of attrition-based warfare. Heavily defended beaches necessitated immense numbers of men and material to capture and hold. No matter how well executed, beaches provided exposed targets to a well-entrenched enemy. The D-Day landings cost the allies ~10,000 casualties.[62] These losses could have been far more horrific, German defenses were less than 20% complete and forces had been deployed elsewhere due to the successful allied deception campaign.[63] The campaigns of Iwo Jima resulted in ~26,000 casualties, Okinawa ~62,000, Peleliu ~10,000. Seventy years later, amphibious landings are not only just as exposed, but weapon systems have improved considerably. A single air strike could wipe clean a full beach. A salvo of missiles or a heavy artillery barrage would generate the same result. A major invasion of North Korea, for instance, would be devastating to the Marine Corps if it attempted to capture the beaches head on and repeat the success of Incheon under heavy fire.

---

[61] For instance: Matt Drake, "More than 200 ISIS militants WIPED OUT in precision Russian air strike in Syria" *Express*, Last Updated August 21, 2017, https://www.express.co.uk/news/world/843778/ISIS-Syria-Fighters-200-killed-Russia-Airforce-Deir-ez-Zor-terror-terrorism-war, Accessed June 9, 2018; Michael Hoffman, "US Air Force Targets and Destroy ISIS HQ Building Using Social media," *Military.com*, June 3, 2015, https://www.military.com/defensetech/2015/06/03/us-air-force-targets-and-destroys-isis-hq-building-using-social-media, Accessed June 9, 2018.
[62] Ken Ford, Steven J. Zaloga, *Overlord: The D-Day Landings* (Oxford, New York: Osprey Publishing, 2009), pp. 335.
[63] Chester Ilmot, *The Struggle For Europe* (Hertfordshire: Wordsworth Editions, 1997), pp. 290.

In order to take advantage of modern small arms and mitigate against the destructiveness of artillery and missiles, small units must be more and more dispersed and take every advantage of terrain possible, which reduces the ability to command and control further (not to mention the improvements of motor vehicles and other mobile assets which increase the speed and scope of operations).[64] In the words of MCDP1, "by historical standards the modern battlefield is particularly disorderly." It is therefore imperative that each subordinate understand his purpose, to take initiative when opportunities arise and to communicate with adjacent units and their command. Independently operating fire team and squads are the bedrock of war in the 21st century, just as the company and platoon were in the 20th century. These trends have not gone unnoticed by the Marine Corps. Every statement made by the current Commandant, General Neller, suggests he is pivoting the Corps to be more innovative, harness emerging technologies, and to employ more capable, older, more experienced Marines.[65] Without these changes, any future conflict is likely to be far more

---

[64] Martin Van Creveld, *Command in War* (Cambridge, MA: Harvard University Press, 1985), pp. 190f.

[65] Lance Bacon, "Commandant Looks to Disruptive Thinkers to Fix Corps Problems," *Marine Corps Times*, March 4, 2016, https://www.marinecorpstimes.com/news/your-marine-corps/2016/03/04/commandant-looks-to-disruptive-thinkers-to-fix-corps-problems/ (Accessed May 9, 2018); Jeff Schogol, "The Next Fight: The Commandant is Pushing the Corps to be Ready for a 'Violent, Violent Fight," *Marine Corps Times*, September 18, 2009, https://www.marinecorpstimes.com/news/your-marine-corps/2017/09/18/the-next-fight-the-commandant-is-pushing-the-corps-to-be-ready-for-a-violent-violent-fight/ (Accessed May 9, 2018); Megan Eckstein, "Interview: CMC Neller Lays Out Path to Future U.S. Marine Corps," *USNI News*, August 9, 2016, https://news.usni.org/2016/08/09/interview-cmc-neller-lays-out-path-to-future-u-s-marine-corps (Accessed May 9, 2018) General Neller, Message to the Force 2018: "Execute." General Neller, Message to the Force 2017: "Seize the Initiative."

devastating. Landing fewer but more effective Marines could mitigate potential deaths, especially from artillery and automatic fire, while increasing the tempo of movement, exemplified by the British landings during the Falkland's War in 1982.[66]

Beyond speed and efficiency, an additional benefit of a decentralized system is its resiliency. It is safe to assume the enemy will try to attack our own centers of gravity and critical vulnerabilities. With distinguishable command and control systems, enemy sabotage, electronic warfare, or artillery attacks could render the unit directionless. A decentralized system would reduce the impact that such a loss would have on the various subordinate units, a trait heavily lauded in *The Starfish and the Spider*.[67]

In conclusion, a brief overview of the necessities for successful command in maneuver warfare suggest that the prescriptions of MCDP1 are congruent with its successful application. It is clear that the only way to implement maneuver warfare is to foster independent thought, self-confidence, and professionalism in all troops. But, how can junior Marines or junior NCOs have this capability if they are trained to be automatons without practice of basic tactics? While the theoretical background to succeed on the battlefield exists, the training apparatus and frame of mind currently does not exist in basic training.

Counterinsurgent/Unconventional Warfare
Having broadly covered conventional warfare, we turn to the demands of unconventional warfare. We noted in

---

[66] Edgar O'Balance, "The Falklands, 1982," in Merrill Bartlett, ed. *Assault from the Sea: Essays on the History of Amphibious Warfare* (Annapolis: Naval Institute Press, 1983), pp. 432f.
[67] Ori Brafman and Rod Beckstrom, *The Starfish and the Spider* (New York: Portfolio, 2006), pp. 21, 39.

Chapter 2 and in Arreguín-Toft's research that the frequency of unconventional warfare has been increasing and will likely not disappear, making training for unconventional warfare perhaps of greater importance than preparing for its conventional counterpart.

As Robert Kaplan writes in *The Revenge of Geography*, "Understanding the map of the twenty-first century means accepting grave contradictions. For while some states become militarily stronger, armed with weapons of mass destruction, others, especially in the Greater Middle East, weaken, they spawn sub-state armies, tied to specific geographies with all of the cultural and religious tradition that entails. Thus, they fight better than state armies on the same territory ever could. Southern Lebanon's Hezbollah, the former Tamil Tigers of northern Sri Lanka, the Maoist Naxalites in eastern and central India, the various pro-Taliban and other Pushtun tribal groupings in northwestern Pakistan, the Taliban itself in Afghanistan, and the plethora of militias in Iraq, especially during the civil war of 2006-2007, are examples of this trend of terrain-specific sub-state land forces. For at a time when precision-guided missiles can destroy a specific house hundreds of miles away, while leaving the adjacent one deliberately undamaged, small groups of turbaned irregulars can use the tortuous features of an intricate mountain landscape to bedevil a superpower."[68]

I have previously suggested that the Marine Corps perform a strategic pivot and take up the role of the nation's primary force for unconventional warfare; this form of conflict must be briefly discussed before moving onto proposed changes in training. The United States has in recent history been successful at fighting

[68] Robert Kaplan, *The Revenge of Geography: What the Map Tells us About Coming Conflicts and the Battle Against Fate*, pp. 125.

conventional wars, but hapless in unconventional campaigns - a history extensively detailed in *Chasing Ghosts: Unconventional Warfare in American History*, by John Tierney. In his conclusions, he notes that the United States has still not comprehended the most basic lessons about guerilla war.[69] Even if the Corps does not perform the strategic pivot suggested in this book, it is still a reality that unconventional wars are increasing in number; training and institutions must adapt to the world they inhabit if they have any hope of being effective.

The exigencies upon a soldier or Marine in an unconventional war are as great, if not greater, than in a conventional conflict. While unconventional wars are increasing in number, the likelihood that conventional powers like the U.S. wins has been decreasing in frequency; it is vital that lessons from past conflicts, both successful and not, be used to shape our understanding of the necessary skills to be developed in modern Marines.

As a preface, in *The Starfish and the Spider*, Brafman and Beckstrom identify three strategies to defeat radically decentralized organizations: (1) destroy their foundation by changing the ideology, which is very difficult; (2) centralize the enemy – such as by introducing property rights and allow power to concentrate[70] – making the now centralized opponent vulnerable and controllable – for example the Apache; (3) decentralize your own organization – as evidenced by terrorist hunting cells in the Middle East.[71] Other than

---

[69] John J. Tierney Jr. *Chasing Ghosts, Unconventional Warfare in American History* (Washington D.C.: Potomac Books, Inc., 2006), pp 260.
[70] When property and property rights were introduced to the Apache, not only did hierarchies develop based on the ownership of resources, but the Apache were less willing to dissolve away when under attack for fear of losing their property – rendering them vulnerable.
[71] Ori Brafman and Rod Beckstrom, *The Starfish and the Spider* (New York: Portfolio, 2006), Chapter 6.

strategy (2), which may not actually function since centralized organizations could decentralize again, and decentralized organizations still thrive where there are already property rights and hierarchies in much of the world in which insurgent combat takes place, both strategy (1) and (3) will reverberate through the prescriptions identified by the counterinsurgent theory presented below.

David Galula's counterinsurgency theories are the bedrock of modern practices. He describes the counterinsurgency war in North Africa (the Algerian War, 1954-1962), the role of the warrior, and tactical considerations in military operations. He writes, "A revolutionary war is primarily a war of infantry. Paradoxically, the less sophisticated the counterinsurgent forces, the better they are. France's NATO divisions were useless in Algeria; their modern equipment had to be left behind, and highly specialized engineer or signal units had to be hurriedly converted into ordinary infantry."[72]

In this type of war, soldiers (and Marines) do not have typical roles to play. "To confine soldiers to purely military functions while urgent and vital tasks have to be done, and nobody else is available to undertake them, would be senseless. The soldier must then be prepared to become a propagandist, a social worker, a civil engineer, a schoolteacher, a nurse, a boy scout."[73] More specifically, Galula argues that soldiers can and should be used in "making a thorough census, enforcing new regulation on movements of persons and goods, informing the population, conducting person-to-person propaganda, gathering intelligence on the insurgent's

[72] David Galula, *Counterinsurgency Warfare: Theory and Practice*, (Westport CT: Greenwood Publishing Group, Inc., 2006), pp. 21.
[73] David Galula, *Counterinsurgency Warfare: Theory and Practice*, pp. 62.

political agents, implementing various economic and social reforms, etc. all these will become their primary activity. They have to be organized, equipped, and supported accordingly."[74] In sum, military forces should be reprogrammed to restore order, providing the basic needs of a society so that insurgent forces lose their civilian base of support – a point echoed by Ricky Waddell in an essay on the application of maneuver warfare to counterinsurgency practices.[75] Likewise, based on one of the only successful counterinsurgent campaigns in recent history, the British experience in Ireland in the 1960's and 70's, Van Creveld argues that counterinsurgent forces should rely less on magnificently armed military troops than on long-term deployments of disciplined, law abiding police forces.[76]

For military operations, which may arise from time to time, Galula says that small independent units must be employed as part of a larger strategy, and should be trained to track guerillas through small-scale operations and ambushes (since guerillas would avoid confrontations that are too open due to conventional overwhelming technology and firepower).[77] Organizationally, "the subdivision should be carried out down to the level of the 'basic unit of counterinsurgency warfare:' the largest unit whose leader is in continuous contact with the population. This is the most important unit in counterinsurgency operations, the level where most of the practical problems arise, where the war is won or lost."[78] Everything prescribed by Galula revolves around carefully crafted policies enacted through small

---

[74] Galula, pp. 66.
[75] Ricky Waddell, 'Maneuver Warfare and Low-Intensity Conflict' in Richard Hooker, ed., *Maneuver Warfare, an Analogy* (Novato, CA: Presidio Press, 1993), pp. 130ff.
[76] Martin Van Creveld, *The Changing Face of War*, pp. 229ff
[77] Galula, pp. 76f
[78] Galula, pp. 78.

units actually present on the ground. Small units and decentralized tactics not only can destroy their opponents and pacify areas over the long run, they can force the enemy to gather a large force – creating conventional battles out of an unconventional conflict, which in turn again can play into the strengths of Western militaries.[79]

Seeking to connect the theory of maneuver warfare and counterinsurgency operations, Ricky Waddell essentially blends the previous discussion of *Auftragstaktik* with Galula's theories. "A counterinsurgency effort in the guerilla phase faces a dual politico-military task. The government must counteract the growing military actions of the insurgents while simultaneously attempting to win back politically the disaffected populace...[80] Since counterguerrilla operations are by nature dispersed, mission orders and decentralized execution should be hallmarks of these military actions. Counterguerrilla actions are not won by big battalions, but by better platoons and squads."[81] The connection between the civilian population and the military cannot be overstated enough. Since military forces must restore order and turn the population against the insurgents, a strong relationship must be forged with local inhabitants. The inability to do so has been a critical vulnerability in the past. "Historically, such weaknesses [of counterguerrilla forces] have been a lack of intelligence, which is often linked to a lack of contacts among the affected population; a tendency to concentrate in formations or locations that become easily attacked targets and a tendency to over-rely on firepower, which

[79] Bing West, *The Village* (New York: Pocket Books, 1973), pp. 229ff.
[80] Ricky L. Waddell, 'Maneuver Warfare and Low-Intensity Conflict,' in Richard Hooker, ed., *Maneuver Warfare, an Analogy* (Novato, CA: Presidio Press, 1993), pp. 134.
[81] Ricky L. Waddell, 'Maneuver Warfare and Low-Intensity Conflict,' pp. 135.

inevitably affects civilian populations to the detriment of the government's attempts to bolster its legitimacy."[82] In practice, forging connections with local civilians requires long term contact with the population to create personal relations, as evidenced by the experiences of Marine 'Combat Action Platoons' in Vietnam. Local soldiers should be able to say that counterinsurgents are "[Marines or Soldiers] who came and lived with us and ate our food and [fought with us] and never took anything from us."[83]

The theory has seeped into field manuals of the Marine Corps, which mirrors the ideas of previous authors. *The U.S. Army and Marine Corps Counterinsurgency Field Manual* is dedicated to arguing that counterinsurgent operations are based on integrating civilians with military personnel, comprehending complex local dynamics, complying with nuanced counterinsurgent rules of engagement which aim to delegitimize insurgents, building basic infrastructure and improving relations with local populations, employing small units to clear and hold areas of hostiles, policing, training allied forces, gathering and utilizing intelligence.[84]

On the topic of command, the manual states: "Local commanders have the best grasp of their situations. Under mission command, they are given access to or control of the resources needed to procure timely intelligence, conduct effective tactical operations, and manage [all direct] operations. Thus, effective [counterinsurgent] operations are decentralized, and higher commanders owe it to their subordinates to push

---

[82] Ibid.
[83] Bing West, *The Village* (New York: Pocket Books, 1973), pp. 213.
[84] U.S. Army Field Manual No. 3-24, Marine Corps Warfighting Publication No. 3-33.5, *The U.S. Army, Marine Corps Counterinsurgency Field Manual* (Chicago: University of Chicago Press, 2007).

as many capabilities as possible down to their level. Mission command encourages the initiative of subordinates and facilitates the learning that must occur at every level. It is a major characteristic of a [counterinsurgent] force that can adapt and react at least as quickly as the insurgents."[85] As Brafman and Beckstrom show, decentralized organizations adapt incredibly quickly. Since insurgents are able to adapt, only radically decentralized counterinsurgent forces may be able to keep adapting at a similar pace. Just looking at the experiences of combat action platoons in Vietnam or counterinsurgent units in Afghanistan, every patrol may require different tactics, every unit may have to deceive the enemy to flush them out, to gather and interpret available intelligence, grasp and implement the theory of being a 'combat hunter,' adopt or adapt new gear and outfits (i.e. "go native"), get along with civilian villagers of a different ethnicity and culture, and scrounge for local materials to construct various buildings.

In summary, counterinsurgent units must be radically decentralized and incredibly adaptive. At the core, this goes against the very foundation of first-generation linear combat. Screaming "EYEBALLS!" at a villager and showing her, you can drill is simply not going to gain her trust or help locate and thoughtfully neutralize a hidden cell of enemies. It is with these concepts of modern war in mind that we must tweak the basic training provided to Marine Recruits and junior Marines.

---

[85] U.S. Army Field Manual No. 3-24, Marine Corps Warfighting Publication No. 3-33.5, *The U.S. Army, Marine Corps Counterinsurgency Field Manual,* pp. 47.

Chapter 4: If it Ain't Broke

The Marine Corps excelled in its duties during much of the 20th and 21st centuries. It performed admirably in almost all recent major conflicts in which the United States was involved in – particularly WWII, the Korean War, and the wars in the Middle East. If that is the case, why does anything need to change in the training regimen that has contributed to the Marine Corps' successes? In colloquial terms, 'if it ain't broke, why fix it?' While current training does a capable job of turning out basic Marines, those Marines are trained for a world in which warfare has now evolved.

When Basic Training was formalized in the early 1900's, the program of first-generation warfare was very similar to that which Marines undergo today, "The course of instruction lasted eight weeks. The first three weeks were devoted to instruction and practice in such activities as close-order drills, physical exercise, swimming, bayonet fighting, personal combat, wall-scaling, rope-climbing, etc. During the fourth and fifth weeks the Recruits perfected their drills, learned something of boxing and wrestling, and were taught interior-guard duties and exercised in extended order. The last three weeks were spent on the rifle range."[1] Boot Camp is now three months long, and activities have

---

[1] Department of the Navy - Headquarters United States Marine Corps, *A Brief History of the Marine Corps Recruit Depot Parris Island, South Carolina 1891 – 1962* (WASHINGTON, D. C.: Historical branch, G-3 division headquarters, U. S. MARINE CORPS, 1962), pp. 4; available online at: http://www.au.af.mil/au/awc/awcgate/usmchist/parris.txt

become testable and part of larger programs (such as MCMAP) – otherwise little has changed. But the Marines needed in the early 20th century are no longer the Marines needed in the 21st. Other than minor modifications, the Marine Corps still trains basic Marines for a pre-WWI world. That deficiency was acceptable in the past because of American material preponderance – but that geopolitical status has also changed. In the last chapter, I discussed the combat situation of the 21st century and Marine Corps doctrine which reflects upon the nature of war. What makes a reformation of training so critical is the impact of long-term trends on American supremacy. I will now briefly discuss the changed geopolitical picture of the world before reflecting upon the optimal use of Marines and their retention rates to highlight additional reasons why training must be modified.

As I previously wrote, the most important cause of American victories in the 20th century were its enormous resources. Before WWII had begun, the U.S. economy was already over twice as large as Germany's, and almost five time as large as Japan's. By 1944, U.S. GDP was almost four times that of Germany and almost eight times that of Japan's. The population of the U.S. was almost as large as the sum of Japan and Germany. With this economy, the U.S. far outproduced its rivals and allies in almost every weapon and combat vehicle, while still placing a smaller burden on its overall economy than the Axis powers.[2] As a result, American troops were superbly equipped, a story repeated through every conflict of the 20th and 21st centuries. For instance, in World War II an American division was equipped with over three times as many supplies as a typical

---

[2] Harrison, Mark, ed., *The Economics of World War II: Six Great Powers in International Comparison* (Cambridge: Cambridge University Press, 1998) pp. 21ff

German division, and over twice as much as a German armored division.[3] In 1939, the U.S. Navy was composed of some 394 vessels, including 15 battleships and 5 fleet carriers. By the end of WWII, that number had jumped to 6,768 ships, including 23 battleships, 28 fleet carriers with another 71 escort carriers.[4] The Navy began the war with about 10,000 aviation personnel and nearly 2,700 pilots – and ended it with almost 440,000 personnel and over 60,000 pilots. Allied amphibious operations against Axis powers were consistently successful because of American supporting arms, especially naval gunfire. A Japanese report estimated that a single American battleship provided as much firepower as five Japanese divisions![5] This superiority continued through the 20th century, As Van Creveld writes about American matériel for Vietnam, "following the American military tradition as it had been since at least the Civil War, the troops came superbly equipped, possessing everything from satellite communication to ice beer, heavy bombers to dancing girls."[6]

These material accomplishments were only possible because of the industry-heavy, comparatively huge economy of the U.S. Contrary to popular conception of the economy broadcast by populists and some mainstream televised media, the overall U.S. economy is not 'getting worse.' Instead, rival economies have 'caught up.' Thus, while the economy has continued to grow, its relative dominance has slipped. In

---

[3] Van Creveld, Face of War, pp. 140.
[4] Naval History and Heritage Command, "US Ship Force Levels: 1886-Present," November 17, 2017,
https://www.history.navy.mil/research/histories/ship-histories/us-ship-force-levels.html
[5] Theodore Gatchel, Defense at Water's Edge: Defending Against the Modern Amphibious Assault (Annapolis: Naval Institute Press, 1996), pp. 142.
[6]Van Creveld, Face of War, pp. 222.

2018, the U.S. economy had a GDP of about $20 Trillion, followed by its rival China with almost $13 Trillion, which had a growth rate around twice that of the U.S., suggesting that by 2035 China's economy will be larger, and approaching parity until then. But China has major advantages the U.S. does not: a population over four times as large, a treasury surplus (instead of major deficits), and a higher manufacturing output.[7] Further, China's economy is mostly based upon industry – both light and heavy - while the American economy has pivoted to be almost entirely service-based. Unlike the U.S., whose obligations spread its forces across the world, China can concentrate its forces in a much smaller area, such as its Southern border, coast, or the South China Sea. A more zoomed out view of long-term trends shows that the U.S. share of global GDP has shrunk over the past half-century. After WWII, the U.S. represented almost 40% of the world GDP whereas now its economy represents closer to 15%.

Without material supremacy, American troops do not always outshine their opponents to the same extent as WWII. In Korea, troops stripped of their massive WWII budget were bulldozed by North Koreans and almost annihilated outside of Pusan, only to be saved by the lucky landing at Incheon.[8] The same story was repeated when China entered the war alongside North Korea. When the impact of technology is reduced in an

[7] David Sims, China Widens Lead as World's Largest Manufacturer, *Thomas,* March 14, 2013, *https*://news.thomasnet.com/imt/2013/03/14/china-widens-lead-as-worlds-largest-manufacturer; China is not alone, India for instance has a GDP of almost $3 Trillion and is the fastest growing economy in the world in part because in half a decade its population is forecast to surpass that of China. Robert Kaplan's *Monsoon* (New York: Random House, 2010) discusses these topics extensively.
[8] Theodore Gatchel, Defense at Water's Edge: Defending Against the Modern Amphibious Assault (Annapolis: Naval Institute Press, 1996), pp. 174.

unconventional war, American troops do not claim decisive victories, as the wars in Vietnam and Afghanistan demonstrate.

Yet, while the material advantages of the US are fading, the story is not all bleak militarily because of the switch to maneuver warfare as a basis for tactics and strategy. As I discussed in previous chapters, Germany was able to defeat far better equipped opponents, and hold its own against overwhelming enemy forces and supplies in both World Wars. In WWI, the German army bolstered all the central powers, fighting successfully with a fraction of the economic power the allies could bring to bear. German tactics were superior to those of their opponents through much of the war. Likewise, in WWII, in just six weeks Germany defeated France, the Netherlands, and Belgium. France alone possessed a larger army, superior and more numerous tanks and artillery pieces than Germany, whose army had only just captured Poland.

Likewise, using tenants of maneuver warfare, Israel continuously defeated coalitions of states, each of which were larger, more populous, and with larger economies than it had – notably in the War of Independence, the Six Day War, and the Yom Kippur War. Maneuver warfare, done correctly, can be the great force multiplier, and it relies on critical thinking, decisiveness, and mental flexibility which can all be developed through education.

While America's material superiority slips, its population has become better educated. In the 1940's, less than 30% of white males had completed 4 years of high school by the time they were 25; for minorities it was a third of that rate.[9] By the 21st century, almost 90%

---

[9] Thomas Snyder, ed. 120 Years of Education: A Statistical Portrait (US Department of Education: Office of Educational Research and

of adults had completed high school.[10] In the 19th century, the average public school student spent less than 80 days in school. Today they spend about twice that, while the pupil to teacher ratio has decreased from almost 40:1 to less than 20:1 over that period.[11] Today, over a million bachelors' degrees are awarded annually, along with hundreds of thousands of masters' degrees, and several tens of thousands of doctorates. More doctorates were awarded in the 21st century than were bachelors' degrees in the 19th century.[12]

Almost all the best universities in the world are located in the United States or in United Kingdom where the individualistic culture is particularly conducive to critical thought and innovation; students from all over the world strive to gain entry to American colleges and graduate schools. As the U.S. became a service-based economy, the education of its population improved considerably. Not only is that important economically, it is necessary for modern war. Exercising the mind is the best way to develop critical thinking, which is the fundamental requirement of maneuver warfare. Indeed, an established conclusion of research in industrial psychology demonstrates that cognitive ability is directly tied to any job performance.[13]

Improvement, 1993), pp. 8; available online at:
https://nces.ed.gov/pubs93/93442.pdf.
[10] Camille L. Ryan and Kurt Bauman, Educational Attainment in the United States: 2015 (U.S. Department of Commerce: Economics and Statistics Administration: U.S. Census Bureau, 2016), pp. 1
[11] Camille L. Ryan and Kurt Bauman, Educational Attainment in the United States: 2015
, pp. 27, 29.[12] Camille L. Ryan and Kurt Bauman, Educational Attainment in the United States: 2015, pp. 75
[13] F.L. Schmidt and J.E. Hunter, "The validity and utility of selection methods in personnel psychology: Practical and theoretical implications of 85 years of research findings." Psychological Bulletin, Vol. 124 No. 2 (1998), 262ff.

Yet in the Enlisted ranks, the boon of education is not being harnessed. Instead, the Corps is hurt by the effects of a more complex and thought demanding economy as individuals choose more intelligent jobs, leaving their inflexible and mechanical tasks behind. Alternatively, those skilled members of the armed forces who enjoy their jobs simply transfer to a private contracting company such as the infamous firm currently called Academi - previously known as Blackwater. Marines are no longer ruffians in need of extreme discipline from the age of press-gangs. In Chapter I, I discussed some of the macro trends affecting new recruits to show that the potential pool is somewhat endangered. I suggested cutting down the number of MOS to reduce the overall number of recruits admitted every year, leaving only room for top-tier applicants. That's only one half of the equation regarding a high quality and experienced Corps, the other half has to do with retention. Retention is essential because older workers are more proficient and effective, which is particularly important in the 21st century as operational needs have become more complicated. Armies before the early industrial era predominately required logistical expertise, a signal corps of some kind, laborers, and various combat personnel. Today, cyberwarfare, electronics, airpower, air defense, and armored units have increased the number of skilled jobs in the military enormously. As General Neller noted, "A lot of the capabilities [Marines are] going to need are going to take more training. It's going to be a longer block of time to grow these Marines, and we're going to have to keep them longer."[14] Experience is necessary for maneuver

---

[14] Gidget Fuentes, "Neller: Future Marine Corps Could be an 'Older, More Experienced' Force," *USNI News*, February 8, 2018, https://news.usni.org/2018/02/08/neller-future-marine-corps-older-experienced-force.

warfare as practice yields better, faster, and more accurate decision making. Even beyond combat, experience affects a host of other factors, from increased productivity, improved readiness, to overall safety. Studies suggest that the increase in productivity as a result of experience alone is large enough to offset the cost of paying higher-ranking service members.[15]

Lastly, a high retention rate improves the overall leadership. If everyone wants to stay in, only the best are allowed to remain as the inferior are weeded out. But in the words of a Sergeant I spoke with about this topic, "It's pretty difficult to kick people out of the Marine Corps because nobody wants to stay in."

The Marine Corps in particular has a major retention issue. Every year, the Corps needs over 30,000 new recruits. In 2019 the Corps will actually require 38,500 new Marines to maintain current numbers.[16] That's a churn rate of over 18% - if we count both Active Duty and Reserve Marines – or over 20% if we only count Active Duty Marines; about a fifth of the Marine Corps departs every year. We find 29% of Marines leaving the Corps each year are junior Marines, 60% are Noncommissioned Officers, and 11% are Senior NCOs. In tandem with this ratio, the majority of Marines leave at the end of their first contract; almost all the rest either depart the Corps at the end of their second contract or at their 20 year mark - which corresponds to their pension-

[15] Jennifer Kavanagh, "Determinants of Productivity for Military Personnel, a Review of Findings on the Contribution of Experience, Training, and Aptitude to Military Performance," *RAND Corporation* (Santa Monica, CA: 1981), pp. 4.
[16] Shawn Snow, "Facing retention issues, the Corps needs to Recruit highest number of Marines in a decade," *Marine Corps Times*, November 5, 2018, https://www.marinecorpstimes.com/news/your-marine-corps/2018/11/05/faced-with-retention-issues-the-corps-needs-to-Recruit-highest-number-of-marines-in-nearly-a-decade/.

funded retirement.[17] Over twice as many Marines depart within their first 4 years - their first contract - than do those who stay for 8 – the end of their second contract. To increase retention and thereby ameliorate the Marine Corps, more Marines need to stay on for at least a second contract.

So why do Marines leave the Marine Corps? Are there any clear solutions? Fortunately, the military already takes retention surveys, which provides us some insight. The survey shows that Marines are proud to be part of the Corps and to train and lead Marines. All other reasons to stay in the Marine Corps are more nuanced. In contrast, the top reasons to leave the Marine Corps are clear with the following order of importance: civilian job opportunities, number of hours worked, housing quality and availability, desire to attend college, pay, and job satisfaction.[18] While housing opportunities and pay are beyond the purview of this book, all other factors could be alleviated by a healthy dose of critical thought at the local level.

Civilian job opportunities are certainly tied to pay, but they are also tied to job satisfaction; the prospect of another job must appear superior to the current condition of being a Marine. Likewise, the number of hours worked is relative to peers or to other potential work. Finally, Active Duty and Reserve Marines can attend college while being Marines. The former have additional perks at their disposal such as Tuition Assistance, while the latter have almost total freedom

---

[17]Aline Quester, Laura Kelley, Cathy Hiatt, Robert Shuford, "Marine Corps Separation Rates: What's Happened Since FY00?," *CNA Analysis and Solutions*, October 2008, pp. 4; available online at: https://www.cna.org/CNA_files/PDF/D0018759.A2.pdf

[18] Headquarters U.S. Marine Corps, "FY 2016 EAS Enlisted Retention Survey Results, July 14, 2016, https://www.hqmc.marines.mil/Portals/61/FY16%20ERS%20Final.pdf?ver=2016-07-14-141912-647, slide 15.

since the cumulative sum of service owed amounts to only about one month a year. Many Marines decide to leave the Corps either because the work of a Marine is not fulfilling or because the demands of the job do not align with the returns.

One of the best ways to improve job satisfaction and make it more efficient is to provide more chances to make decisions at a lower level. Allowing and incentivizing Marines to make thoughtful decisions would go a long way in increasing job satisfaction. Instead of implementing the same top-down process day in and day out, Marines of all ranks – particularly the ones doing the implementation – should strive to ameliorate their condition by developing new procedures and technologies. That is in fact one of the primary sources of innovation in economic theory, notably espoused by Adam Smith. Given the relationship between adaptation, maneuver warfare, and success in war, it is shocking the military does not harness the innovative potential of its members.

Appearances, conformity, and standards are prized above everything else, and that's a terrible shame. Institutionalizing critical thought at all levels would cut away at the shallow corporate environment that has been created in today's Marine Corps. Garrison procedures provide a salient set of examples of the problems created by a culture devoid of critical thought. Field day, punishments, timeliness, and weekend liberty briefs are institutionalized procedures that highlight wasteful shortcomings.

Every week most junior Marines spend hours cleaning their rooms à outrance during field day. Originally, field day meant that Marines returning from the field cleaned their gear, workspace, and living quarters before enjoying the weekend. The intent is to have a clean living space; the implementation is often overboard. NCOs spend as much or more time on Field Day than their subordinates inspecting and supervising to ensure that living quarters are more spotless than a five-star hotel. Though cleanliness is paramount in the military, mostly because of the devastating effects of disease, to decide that an otherwise clean room is dirty because dust was found in a corner of a door hinge or because a hair was hiding in the grout of restroom tiles results in an enormous waste of time for both junior Marines and NCOs, and makes life in the Corps extremely disagreeable.

Marines clean for hours during the day and evening, and again in the morning the day before inspection. Those that fail clean again, often cutting into the workday and sleeping hours. A major unintended consequence is that Marines are more likely to enter a hasty marriage just to have the option to live out of the barracks – which in turn contributes to the comparatively high rate of marriage and divorce in the Marine Corps; the marriage rate of young Marines is three times that of their civilian counterparts and, in this population group, the divorce rate is also three times greater than in the civilian world.[19]

Whenever a Marine makes a mistake in society, the entire unit is punished. For instance, if an individual got into a drunken fight in a bar, they are not only punished in the civilian world. They are also severely penalized in the military – oftentimes their careers are ended. However the punishments do not end there, the leaders of the unit are also held responsible for that Marine, the NCOs might also receive negative paperwork, the Officer may be relieved, and the rest of the unit is punished – typically with "games" adapted

---

[19] Michael Foskett, "The Impact of Divorce Among Marines, E-5 and Below, on Unit Operational Readiness," *USMC Command and Staff College: Marine Corps University*, December 4, 2013, Appendix A.

from Boot Camp such as constant mass formations to 'ensure everyone is accounted for,' cleaning without end, or keeping Marines from going to sleep. The idea is that the unit failed to keep that Marine accountable or did not provide them with proper training – but oftentimes the unit had no idea what that individual was up to. Marines are both people and warriors. It is insane to think that people in general won't make stupid mistakes, and even more so if they have been trained to be aggressive risk-taking fighters. Punishing 99% of a unit because 1% made a mistake is terribly wasteful and frustrating. This dramatically increases dissatisfaction. My point is that there will always be a margin of error and no group policy will ever overcome that margin, so instead of pretending that it can, a more optimal policy is to punish individuals as the situation dictates.

If it hasn't been made clear, superiors are often leery of their Marines, they don't trust them. A salient impact is that whenever there is a time set for an event, the time at which most Marines are required to be there is pushed back through the ranks. For instance, a Company Commander may decide she wishes to have a run begin at the reasonable time of 0630 (6:30 AM). To make sure everyone arrives on time, the Company Gunnery Sergeant will decide that everyone should show up at 0615. The section leaders will want to make sure that everyone is accounted for before that, at 0600, and so on. Marines will therefore be formed up at 0500 for a run set to begin at 0630. Almost every event or action is set up that way which not only damages morale but wastes significant time.

The defensive policymaking that has infected the Marine Corps can also be found in the weekend brief. Every Friday, most Marine units receive a safety brief for up to half the workday. This "brief" consists of what one is not supposed to do over the weekend:

"Don't get a tattoo without letting your NCO know what it'll be. Don't do drugs. Don't huff paint. Don't drink too much. Don't drink underage. If you're of age, don't drink and drive. Don't add to or subtract from the population. Don't be on the news..." And so on. Potentially every level of leadership – Officers to Staff NCOs to NCOs gives a safety brief. Why? Because once a Marine does something wrong, their superior can claim that they already told them they shouldn't do that and so thereby provided necessary instructions. But the brief doesn't really accomplish anything – no statistical study has been performed to determine its effectiveness; if Marines were going to do something wrong, the safety brief doesn't just pop back into their minds and convince them to be smart. It's an enormous waste of time that carries massive costs; if half of every Friday is dedicated to the brief, that's 10% of the workweek. Seeing that about $13 Billion are spent on personnel wages in the Marine Corps, if most of the Marines in a unit are engaged in giving or receiving the brief, the cost every year is worth around $1 Billion to the taxpayer. Consider that almost everything is micromanaged in such a way and it's easy to see how the enormous budget allocated to the U.S. military is always just not enough.

Why are Marines not trustworthy in the first place? Because they've been micromanaged for everything, they have explicitly been taught to not think for themselves because someone else planned everything. If they were trained to be responsible, and were held accountable for their own actions, they would in turn be more trustworthy; from the moment they join the Marine Corps those concepts need to be drilled into their heads, rather than thoughtless discipline. Ernst Schumacher, the great economist and statistician wrote: "Any intelligent fool can make things bigger, more complex, and more violent. It takes a touch of genius –

and a lot of courage to move in the opposite direction." The Marine Corps is filled with intelligent fools. Whenever something needs to happen, or something goes wrong, the answer has been to develop mass procedures. Deluges of orders and regulations are added every year. Do these prescriptions really make the Marine Corps a more lethal and capable fighting force? Do they contribute to the capture of forward operating bases and amphibious assaults? Not one bit, they're wasteful of time and resources. No wonder Marines are happy to leave mindless task-based work for the freedom of school and more intelligent jobs. Yet applying mission intent thinking (a concept covered later in this chapter) to lower levels would provide the freedom for Marines to actually apply themselves, and in turn gain fulfillment. The first step in bringing cognition to the Marine Corps is to introduce it in Basic Training and to reward it whenever possible, at all levels.

There are two simple solutions to many of these issues:

1. Make Marines more trustworthy.
2. Hold individuals accountable for their actions.

Both are in fact prescribed by the leadership principles "develop a sense of responsibility in your subordinates" and "seek responsibility and take responsibility for your actions," respectively. Though they may be simple solutions, they are not easy to implement because they go against the norms of the 21st Century micromanaged Marine Corps. Basic Training must be modified because it hasn't been truly changed since the early 20th Century. It no longer prepares Marines for modern combat, it is disconnected from modern military doctrine, it does not harness the minds of the countless educated Americans that join, and it ensures that rote tasks are the norm for most Enlisted – thereby affecting job satisfaction and lowering the available experience to defend the nation. While changing training would affect individual Marines (solution #1), the culture and system of leadership must also be adapted to reflect the changes (solution #2).

Leadership at the Lowest Level

The Marine Corps markets itself by stating that leadership is always pushed down to the lowest level, especially compared to the Army. Yet as much as this claim is made, and as true as it might be relative to the Army, in an absolute sense it is simply not the case. While doctrine states that Marines should constantly seek to take initiative and lead, they are not taught to think critically, to exercise judgment, or try to improve conditions and methods. In order to accomplish maneuver warfare, the most important aspect is the mission intent, which is included in a five-paragraph order. The general components of a five-paragraph order are: Situation, Mission, Execution (the plan and subtasks), Admin and Logistics (matériel, and administrative matters), and Command and Signal (hierarchy and communications). Commander's intent is typically placed within the 'Execution' portion of the order, though it can also be mixed into the 'Mission' segment. Commander's intent provides context for an action and the general goal, the general "what and why" the action should be done, while the unit is supposed to decide on the "with who and how" the action shall be performed.

For illustrative purposes, imagine a unit with the mission to blow a bridge in order to delay enemy reinforcements should they appear; the commander's intent is that an allied beachhead is vulnerable and needs to be defended, this bridge should be captured because it provides potential access to the beachhead. When the unit arrives at the bridge, local circumstances feature elements of the enemy in full retreat. The unit is positioned to eliminate or capture them. Moreover, scouts report no reinforcements for days. Would it be better to blow up the bridge – which could in turn be used to pursue the foe – or concentrate on destroying the enemy, and consider whether or not destroying the bridge is still advantageous? Because of the context established by the mission intent, neutralizing the enemy becomes the proper action, in spite of the actual order (to blow up the bridge). Critical thinking would have yielded a local victory and a secure crossing point to continue the offensive; lack thereof would have allowed the foe to slip away and delayed any pursuit.

Yet taking initiative beyond one's immediate orders is not actually taught or encouraged at the lowest level, despite the theory the Marine Corps is supposed to subscribe to. Many Marines – especially Enlisted – don't even know what commander or mission intent even is. As my Senior Drill Instructor responded when I asked him why we did certain things, "I don't ask, I just do."

Now there is something to be said for 'doing things by the book,' of following a prescribed system. As commanders have many sources of information, they may be able to aggregate data and thus have better judgment and information than their subordinates, while keeping in mind the big picture. This information is combined into a 'system,' or standard procedure. Because of the information and experience used to develop this system, it is safe, and it covers many conceivable possibilities at the time of creation.

Following the system provides security, both for general operations but especially for one's job. For instance, a procedure might state 'in the event of an attack, immediately don your protective equipment.' If a failure occurs in an operation, the command will investigate whether the prescriptions were followed. If they were not, then the infraction was obviously the cause of the flop! It doesn't matter if it was truly the cause, the fact that the prescriptions were not followed means the individual failed. Failure is thus caused by negligence. On the flip side, an individual who had followed prescriptions and failed would not be criticized. In this case, it was a fault with the system, which could be changed if overwhelming evidence suggests it is broken. But such a change is slow, and systems cannot in any case cover every possibility in an ever-changing world, and especially not on an infinitely complex and changing battlefield. The determinant of success should not be whether all the t's were crossed, and i's were dotted, but whether the mission was accomplished in an effective and efficient manner.[20]

In the example provided above of donning protective equipment, perhaps in some instances it would be more advantageous to grab a weapon and immediately join beleaguered defensive forces in combat; perhaps retreat is a better decision – but the weight of the equipment makes retreat slower, allowing the enemy to catch up and eliminate you.

Marines are currently taught to follow orders directly, without question – in spite of the doctrinal publications. They do not question, and there is almost no room for interpretation. For instance, there are orders on many Marine bases that do not allow individuals to walk or run around with earbuds on to listen to music. The reason for those orders is that wearing earbuds and listening to music is both unseemly and unsafe – the latter because music distracts, increasing the chance that one will be hit by a vehicle or some other accident may occur. But what if, on the same base, an individual ran in

---

[20] This is the main point made in Elbert Hubbard's *A Message to Garcia*, one of the books prescribed to Junior Marines on the Commandant's Reading List.

the hills at night, without anyone around, where there are no cars or people. In theory and practice, military police would pull that individual over and tell them they are in violation of the order. That's nonsense. The runner is not a visual stain on the Marine Corps when their audience is the stars and the sky. What's more, they clearly would not be in danger of being hit were there no cars in the area. The world is not black and white, there is almost always room for interpretation. Instead of trying to accomplish missions, subordinates become more focused on following the scriptures of the system handed to them. Furthermore, a procedure inevitably has weak points. Once an enemy understands a method, it becomes easy for them to overcome it.

An analogy may elucidate the point I'm trying to make. Consider a sports team and their coach. During a game, would it not be infinitely more efficient and effective for each player to know what they are doing and what the aim of the game is (to score), than for the coach to somehow constantly tell each player how to act – which is obviously difficult or impossible considering the number of moving pieces. If that coach wrote exact procedures for a game, and the players followed those procedures, then the opposite team would, over time, learn the various memorized schemes and how to nullify them. In addition, it is impossible for those procedures to account for every possibility. Perhaps an area on the field becomes unavailable, a pass cannot be completed, no one is open to receive, etc. That's why good coaches don't micromanage their teams. They focus on improving their players, ameliorating their team and its cohesiveness, and working on a few general strategies which are inevitably shaped by the idiosyncratic course of the game.

Beyond being inefficient, centralized decision-making bodies are also prone to elitism and what social psychologist Irving Janis called 'groupthink.' It is easy to imagine that the upper echelon of the armed forces of the world's most powerful nation might feel superior to the outside world, while groupthink happens when individual members of the group attempt to minimize conflict within the group by not challenging any seeming group consensus. The term groupthink was in fact invented to explain American policy makers' outrageous decisions that led to the Bay of Pigs and Vietnam military strategy. In fact, in the sphere of military operations, there is no better display of centralized, rule-based management than the conduct of the Vietnam War under Robert McNamara. [21] It was a complete disaster. A decentralized and adaptive force may not entirely avoid the effects of the hubris of elitism or groupthink, but the effects would be constrained to a local unit and not upon the entire military establishment or nation. Yet, some five decades after, the focus on centralized command and control – in spite of the official line of doctrinal publications – is just as strong. Why is that?

The Marine Corps has two conflicting theories of leadership. On the one hand, the Corps prides itself on 'leadership at the lowest level.' This implies that NCOs – Corporals and above – and Officers have the ability to act independently. On the other hand, the Corps also believes that responsibility ultimately rests with the leader. Further, the leadership principal "seek responsibility and take responsibility for your actions" suggests that Officers, SNCOs, and NCOs are incentivized to progressively remove independence from each of their subordinates in order to effectively take responsibility for both their actions and for the actions of those under their command. Not doing so can result in

---

[21] See Deborah Shapley, *Promise and Power, the Life and Times of Robert McNamara* (Boston: Little, Brown and Company, 1993), Part III and IV in particular.

negative consequences for Officers, as a few recent cases show.[22]

It is a bit of a redundant task because no one individual can effectively manage the actions of all individuals under their command; no Officer can eavesdrop on subordinates spread across myriad locations at all times of the day and night. While micromanaging does ensure that tasks are done to standards, it is not correlated with adaptation, is bad for morale, and is exhausting to all parties involved. More importantly perhaps, in battle micromanaging and leadership from the top has disastrous results, exemplified by the capabilities and efficiency of armies using maneuver warfare (i.e. Germany, Israel) compared to centralized, attrition-based armies (i.e. Russia, Arab States) in 20th Century wars.

Doctrinal publications suggest that implementation of mission intent should be at the lowest level; those same publications should add reasonable responsibility to that same level. In training, leadership must balance safety with realism and stress. The issue is that we seem to be too close to safety, and too far from realism. There is no obvious answer for this. It is however absolutely necessary that both the NCOs and junior Marines implementing policies, rather than leaders, be held responsible for their own actions without necessarily impacting the entire chain of command

---

[22] Matthew Cox, "Former Recruit Commander Pleads Guilty in Parris Island Hazing Scandal," *Marine Corps Times*, March 12, 2018. https://www.military.com/daily-news/2018/03/12/former-Recruit-commander-pleads-guilty-parris-island-hazing-scandal.html, Accessed May 5, 2018; Hope Hodge Seck, "Marine Corps Osprey Squadron Commander in Pacific Fired," *Marine Corps Times,* February 1, 2018, https://www.military.com/daily-news/2018/02/01/marine-corps-osprey-squadron-commander-pacific-fired.html, Accessed May 5, 2018.

should that chain not be aware of those actions in order to break the need to supervise and control.

Officers should set policy, delegate responsibilities to their subordinates, and ensure (to the best of their abilities) that the mission is being accomplished effectively. The Marine Corps needs to stop shunting blame into higher echelons and simply hold those who fail accountable. As long as a commander acts properly and tells their Marines what the laws and orders are, their careers should not be ruined if their subordinates did not follow those laws and orders. On the flip side, unthinking subordinates should also be punished for poor decision-making; it seems fundamentally wrong to me that only Lieutenant Calley was convicted for the massacre of My Lai (1968), and that no Enlisted were even charged. A reasonable person and information standard must be applied: a decision made with as much information as available without the benefit of hindsight based upon rationality and intuition should be defended, *even if the consequences are not entirely positive.* Tweaking the locus of complete responsibility will provide more leeway to delegate responsibility and leadership.

To teach *Auftragstaktik* in the German army, Officers went to academies where they were taught to be self-confident, think independently, and disobey orders as long as they could argue those orders mistaken or wrong.[23] Today in the Marine Corps, some Officers may be taught that way, but NCOs and junior Marines definitely are not. One is infinitely more likely to hear an NCO screaming, "discipline is instant and willful obedience to all orders" to their Marines than, "think for

---

[23] Thomas Ricks, "An elusive command philosophy and a different command culture," *Foreign Policy*, September 9, 2011, https://foreignpolicy.com/2011/09/09/an-elusive-command-philosophy-and-a-different-command-culture/

yourself." In fact, I have never heard an NCO tell their Marines that critical thinking is the foundation of modern combat. The draconian discipline in the Enlisted ranks is an enormous weakness in modern combat. The Marine Corps has developed a no-risk mentality, a defensive culture. It provides little incentive to take calculated risks, but nearly infinite criticism when things go wrong.[24] Among companies, research shows that the lower down a manager is in the corporate hierarchy, the more likely she or he is to engage in defensive decision-making, often at the expense of the company's interest. Why? Because they have to in order to "cover their asses."[25]

I am advocating a stronger division between Officers and NCOs where Officers make policies and establish goals and are held responsible for making good policies and goals – like lawmakers. NCOs are responsible for achieving those policies and goals - with some flexibility as to how they wish to implement them - both in combat and not. As I have argued in this book, the modern and future battlefields require more and more dispersion, requiring smarter, more capable Marines; the training should be modified to reflect that NCOs must be capable of making decisions crucial to implementing maneuver warfare. In turn, junior Marines must be taught the fundamentals of leadership, decision making, and critical thinking in order to set them up to either take a leadership position should their immediate superior fall or to be better prepared when they are promoted. Respect for superiors must be based upon their proficiency rather than forced through discipline.

---

[24] Thomas Ricks, "We're getting out of the Marines because we wanted to be part of an elite force," *Foreign Policy*, January 4, 2013, https://foreignpolicy.com/2017/12/20/were-getting-out-of-the-marines-because-we-wanted-to-be-part-of-an-elite-force-4/

[25] Gerd Gigerenzer, *Risk Savvy: How to Make Good Decisions* (New York: Penguin Books, 2015), pp. 114.

I have already discussed the importance of pushing responsibility down to the local level – away from Officers – in order to incentivize decentralization. So, if NCOs are given latitude to make decisions, would they make worse ones than Officers? Is there any reason they should be trusted less than their superiors? If so, on what grounds: what makes Officers superior to NCOs?

First, we must define what services an Officer provides. To quote a RAND research paper prepared for the Air Force, an Officer is "responsible for organizing, leading, and controlling the activities or some phase of the activities of a military organization. [Officers] in combat [are] peculiarly expert at directing the application of violence under certain prescribed conditions and it must be remembered that the skill of the Officer is the management of violence not the act itself."[26] Officers do not directly manage their subordinates, they do so through their Noncommissioned Officers. Junior Officers – Second and First Lieutenants – direct Corporals and Sergeants, who in turn manage Junior Marines. About 19% of the Enlisted Marine Corps is composed of Corporals and 16% of Sergeants for a total of about 60,000 Marines.[27] They are in turn directed by about 6,000 Lieutenants, a ratio of 10:1.[28] Maneuverability and flexibility would increase exponentially if decisions beyond strategy and policy

---

[26] James Hayes, The Evolution of Military Officer Personnel Management Policies: A Preliminary Study with Parallels from Industry, *Rand Corporation* (Santa Monica, CA: 1978), https://www.rand.org/content/dam/rand/pubs/reports/2006/R2276.pdf, pp.1

[27] Headquarters United States Marine Corps, *U.S. Marine Corps: Concepts & Programs 2013, America's Expeditionary Force in Readiness* (2013), Chapter 4, pp. 234

[28] Headquarters United States Marine Corps, *U.S. Marine Corps: Concepts & Programs 2013, America's Expeditionary Force in Readiness*, pp. 230.

were thus made by NCOs, rather than decided upon by Officers, then filtered down onto and through NCOs.

There are two major differences between Marine Officers and Enlisted personnel, which I discussed briefly in the introduction. First, most Officers go through OCS and TBS, Officer Candidate School and The Basic School (respectively), where individuals are evaluated and unsatisfactory candidates are filtered out.[29] Second, Officers are required to graduate college. In this section, I argue that there is no reason why junior Officers will automatically make better tactical decisions than their Noncommissioned Officers. In other words, going through OCS, TBS, and college does not necessarily make an Officer a better manager or decision maker than an NCO.

In theory, a newly minted Officer is supposed to have more life experience than the Marines they lead. In practice however, that is far from the case. A college graduate of twenty-one or twenty-two years of age leads Corporals and Sergeants of the same or greater age. Almost 50% of recruits are between 17-18 years old and about 35% are between 19-20 (the rest are older). Before becoming eligible to become a Noncommissioned Officer, newly Enlisted Marines must first spend four months in Basic Training (Boot Camp and MCT), and about another half a year to become a Lance Corporal – the highest rank a junior Marine can be. Several additional months can be tacked onto that if the individual in question entered their contract as a Private and not as Private First Class. In total, that's about a year (14 months). Thus, even the youngest Lance Corporal would be between 18-19 years old. Promotion to Corporal is competitive within the job field; in the electronics maintenance occupational field (5900) all

---

[29] I say most because those who attend Annapolis Naval Academy for college do not need to go through OCS.

Junior Marines in their specific MOS compete against themselves. 5939 – Aviation Communication System Technicians - Lance Corporals compete against other 5939's. 5974 – Tactical Data Systems Administrators - compete against each other, and so on in each occupational field and specialty.

To standardize the process, a score is compiled predominantly from one's physical proficiency (based on physical fitness tests), rifle qualification scores, extra-military education, proficiency and conduct marks (a sort of GPA reflecting whether one is a good Marine and good at one's job), time in grade, and time in service. The minimum time in service to be promoted to Corporal is 12 months, with at least 8 months as a Lance Corporal. As a result of the time requirements, even the most brilliant and effective Corporals are at least 20 years old. For a Sergeant, the requirement is 24 months in service and 12 months as a Corporal; even the most capable Sergeants are thus at least 21 years old – though typically Sergeants have between 5-8 years of experience. Rising higher than Sergeant becomes more competitive due to the sudden drop in the number of spots available and because competition escalates from the specific MOS to the general MOS – for example all 5900's compete against one another, 5974's compete with themselves, 5939's, 5979's, 5948's, and so on. While there are about 28,000 Sergeants, there are only about 16,000 Staff Sergeants, and 9,000 Gunnery Sergeants. It becomes very difficult to move up in rank and individuals occupy their positions for longer periods of time. The move from Sergeant to Staff Sergeant is also a major shift in tasks, from active to more passive leadership. Many Marines enjoy hands-on work and the shift to a world of paperwork and indirect management is sometimes not sought. As a result, many individuals stay as Sergeants for several years – until the end of their second contract

- thereby spending about half a decade as Sergeants. Thus, upon graduation from college and Basic Training, Officers do not have immensely more life experience than their NCOs. Nor, in fact, does their life experience necessarily correlate to military or combat matters since that time was predominantly spent in a college environment, which many would say does not provide realistic life experience. In contrast, NCOs have already spent several years in the Marine Corps, thereby garnering more applicable knowledge than their commanders; this dichotomy appears a paradox given that "Knowledge" is an official Leadership Trait, while "Education" or "College" is not. Furthermore, research shows that leaders – particularly those making decisions with little information and time – operate on rules-of-thumb that are in turn based on experience.[30] NCOs with years of experience in the military should thus be better decision makers than their Officers. So, if life experience is not a major point of difference, perhaps secondary school is.

Higher education is important, but does it inevitably improve the pool of newly minted Officers? In the early 20th century, a college degree was rare. A degree was proof that one was a part of the nation's financial and intellectual elite. In 21st century America however, a degree is no longer a major determinant of any kind of special capability. In 2017, about 67% of high school graduates ages 16 to 24 were enrolled in colleges or Universities. Today, over 30% of people over the age of 25 have a bachelor's degree in the US, whereas in the early 20th century only about 5% of the population held a bachelor's degree. The overall average acceptance

---

[30] Gerd Gigerenzer, *Risk Savvy: How to Make Good Decisions* (New York: Penguin Books, 2015), pp. 115

rate for College in the U.S. was 66%;[31] though some colleges have acceptance rates near or at 100%, without even citing for-profit colleges that are incentivized to admit all applicants.[32] Likewise the average graduation rate was almost two thirds, at 60%.[33]

This last percentage hides the fact that college has never been so easy. As the population of individuals attending college has grown, standards have fallen. Students now study less than half as much as their colleges and universities say is required; in the 1960's an average full-time student studied for about 24 hours a week, while today the average is 14 hours.[34] Because of steady grade inflation across the country, excellent grades are now the norm, not to mention the ease with which one can earn a passing score.[35] Most students who fail in college do so not because they failed academically but because they did not have the motivation to continue, or they refused to develop the most basic study habits necessary to pass.[36] In other words, as long as one has the money to continue attending an average college, all one essentially has to do to graduate is wake up in time for a few classes. Bachelor's degrees simply do not

---

[31] Melissa Clinedinst and Anna-Maria Koranteng, 2017 State of College Admission, *National Association for College Admission Counseling*, https://www.nacacnet.org/news--publications/publications/state-of-college-admission/soca-chapter1/.

[32] CollegeSimply has a page dedicated solely to colleges with high acceptance rates, see "U.S. Colleges Easy to Get Into: Colleges with the Highest Acceptance Rate in America for 2018," https://www.collegesimply.com/guides/high-acceptance-rate/?view=all.

[33] National Center for Education Statistics, "Undergraduate Retention and Graduation Rates," Last Updated: May 2018, https://nces.ed.gov/programs/coe/indicator_ctr.asp

[34] Philip Babcock and Mindy Marks, "Leisure College, USA: The Decline in Student Study Time," *American Enterprise Institute*, August 5, 2010.

[35] Stuart Rojstaczer & Christopher Healy, "Where A Is Ordinary: The Evolution of American College and University Grading,1940–2009," *Teachers College Record*, 2012.

[36] Abour H. Cherif, et. al. Why Do Students Fail? Faculty's Perspective, Higher Learning Commission, 2014 collection of papers, 2014.

convey the same special meaning they once did. How then can simply having a degree be a fundamental and major baseline difference between an NCO and an Officer? Without any GPA cutoffs, course requirements, or college rankings, having a degree cannot truly be said to contribute much to "organizing, leading, and controlling the activities or some phase of the activities of a military organization." In any case, the generous provisions provided by the government for education and the promotion incentives related to education ensure than many NCOs are able to educate themselves while serving, even if they have not yet earned a degree.[37]

The last salient difference is the basic training provided to Officers compared to Enlisted. Certainly, the difference between OCS and Boot Camp is stark. However, the changes I propose in this book bridges much of the divide; small unit tactics must be taught at the earliest possible level along with the basic elements of military leadership. Beyond these major facets however, there is little difference – Officers are not taught to think flexibly, they pass OCS if they lead in a specific military manner, as aggressive Platoon Sergeants, otherwise they are dropped for 'failure to adapt.' If anything, the style is infinitely less flexible than those available to NCOs when they manage and supervise their Marines. The 6 months of Basic School, which Officers attend after OCS, is really the time when they practice implementing organizational, management, and leadership skills. Even if the experience of Noncommissioned Officers actively serving with their units does not compare, they attend

---

[37] I realize this argument appears to run counter to my previous claim that the population is becoming more educated – but that is not my intention. I aim to show that the requirement of a degree cannot be assumed to lead to better leaders or thinkers, which does not necessarily mean that the average population is not becoming more educated.

military professional courses to bolster their knowledge: Corporal's Course and Sergeant's Course. Each of these courses are both about a month long and focus on building and reinforcing the fundamentals of basic military leadership. In sum, a Sergeant will have had at least half a year of formal military training in various schools from Boot Camp through Sergeant's course. By that point, an NCO will not only have reached parity with their Officers in terms of formal military schooling but will also have served several years.

If NCOs are currently unable to generate tactical decisions as well as their commanders, it's because they've been taught not to think. Had the Marine Corps focused on building capable decision-makers during its formal training and followed its own operational prescriptions, NCOs would be just as capable of making quality decisions as their leaders. My proposed changes to MCT and Boot Camp would be an enormous step in that direction. Maneuver warfare should be taught as low down as possible. Fluid decision making cannot entirely be the purview of Officers, they should be the purview of everyone; for as I have argued in this book, the modern battlefield requires smaller, more flexible, and more adaptive units than ever before.

Nathan Decety

Chapter 5: Training Changes in Boot Camp in Practice

What should training be like?
In my view it is possible to keep the good elements of basic training while making more productive use of time and resources. The following proposals revolve around making smaller platoons, teaching leadership at the lowest level through experience, forcing more responsibility onto Recruits, compressing some events currently completed in MCT into Boot Camp, and training Recruits in basic infantry tactics. The events from MCT would allow combat training to be more aggressive and tailored to the individual.

The most important issues identified at the heart of the reform are: the platoons spending so much time playing "games," the inefficiencies inherent in unnaturally large units, little repercussions for poor performance, and lack of leadership development for the vast majority of Recruits. Having provided my experiences in Boot Camp in Part 2, the reader may identify similar areas for improvement and many such redundancies. By reducing the size of the platoon, removing unmotivated or underperforming Recruits, and creating opportunities to develop leadership and responsibility, the end result would be a greatly superior average basic Marine.

Platoon Size
My platoon began with 90 individuals. Part of the reason we did so much drill and played so many 'games' is

because of the proverb 'a team is only as strong as its weakest player,' or 'a chain is only as strong as its weakest link.' By the end of the three-month long Boot Camp, there were still individuals who were slow when doing simple tasks and incapable of following simple movements in drill. Because of these Recruits, the platoon was limited in how far it could go.

A smaller platoon would mean a much faster pace. Every activity would be more rapid: from processing through the chow hall to getting into formation for drill. Platoons at OCS were not 90-man strong; they began around 50 or 60. Platoons in the fleet Marine force are generally never larger than 50 Marines. The main impediment to this would be the lack of suitable Drill Instructors: a training regiment of 540 split into 6 platoons of 90 requires about 18 Drill Instructors at minimum for the platoons – 3 Instructors per platoon. If 540 Recruits split into platoons of 60 that would require at least 27 Drill Instructors.

However, recall in that in Chapter 1, I argued that the list of MOS is too great given the requirements of the Marine Corps. Major MOS like administration and supply should be handed off to the Navy, and some of the others should be consolidated. Increasing the efficiency of the Corps would reduce the necessary number of Recruits, meaning the number of necessary Drill Instructors need not change. Indeed, the pressure on individual Drill Instructors would be reduced as they would have to manage fewer Recruits. With a more advantageous ratio of Instructor to Recruit, Instructors would be able to focus on individual Recruits more. No one would 'slip through the cracks,' and feedback could be tailored to individual Recruits. The life of a Drill Instructor is incredibly difficult, and many return to the fleet weary. Improving their conditions would in turn ameliorate the training they provide and reduce the wear and tear they experience.

Selectivity

At every step of recruit training, Instructors would threaten us with being dropped. The procedure of 'getting dropped' is different between OCS and Boot Camp. At OCS, when one was dropped from training, one returned home almost immediately. The individual was not fit to be an Officer. In recruit training, a large number were simply sent back and recycled through a different platoon. Recruits were incentivized to behave well in order to not spend more time in Boot Camp. It's rare for a Recruit to get kicked out fully, a Recruit just goes through more mind-numbing training as a punishment for having been unsuitable in a previous platoon. For those who are fully dropped, many receive severe penalties such as dishonorable or other than honorable discharges in their military records, which negatively impacts them for the rest of their lives. In practice, many Recruits make it through training without passing myriad events and without getting separated from the Corps. Perhaps they had lower proficiency marks on their records than their peers – but those marks affect only promotion and are in any case updated several times – meaning the initial proficiency marks from Boot Camp become averaged away. I think that's a major mistake. A poor performing Recruit who is unable to master the basics of being a Marine is *not* a well-rounded Marine. These individuals are passed through, again, because the Marine Corps needs those Marines to staff its ranks and the organization cannot afford to wait for more qualified applicants.

Boot Camp should be made more like OCS where Recruits can be sent home without a stain on their record. Clearly, the Corps just wasn't for them. Moreover, passing events needs to have a direct impact on being able to continue training. After the first PFT at OCS, several candidates were immediately sent home for being too weak. The same should apply for the IST done at Recruit training, and other events like going through the obstacle course in a set amount of time. Assuming the platoon is only as strong as its weakest member, then if we can eliminate the weak, the average performance will increase considerably. A positive potential consequence is that Recruiters may force their applicants to become more fit on their own so as to not get dropped immediately, thus sending better potential Recruits to Boot Camp, thereby possibly increasing its difficulty. While some may feel this is harsh, Recruits are preparing for situations where performance is a matter of life and death for themselves, their team, and possibly countless others since Marines are the custodians of our nation's safety.

At combat training (MCT), an NCO representing the nearby Recon unit came to speak to us. I vividly recall him saying, "Do you ever feel like training in the Marine Corps is a check in the box?" He was exactly right. There are very few repercussions for being an under average Recruit except that one might be IT'd at a greater rate than the others. Instead, major events should be carefully monitored, and the performance of individual Recruits recorded. Recruits who cannot perform up to standards should be warned, then separated without rancor.

There's another statement we hear a lot as Enlisted, "There's always that 10%," which suggests that about 10% of every unit is useless. That 10% doesn't do their jobs; they can't perform as Marines. Why can we not get rid of the 10% before they become Marines? Increasing the ability to drop Recruits would incentivize better performance and ensure that weak links do not contaminate their units later in the fleet. The average performance of the platoon would thereby increase. Coupled with a reduced size, the platoon would be able to undergo more training events at a more intensive pace.

In the current state, letting Recruits graduate Boot Camp when they underperformed or were unmotivated to become Marines means they will be poisonous assets for the duration of their contract. Rather than just the lost cost of the training they've gone through, the Corps will also pay for at least four years of unsatisfactory performance. Their performance may reverberate to affect their lives, critical equipment, and the lives of other Marines. A Marine who can't defend their fighting hole, or carry their own load, or do a job under pressure could have terrible ramifications during combat. The first and most fundamental index for a Marine should be whether they can function as a rifleman. If they cannot, there is no need to qualify them for additional jobs as they are not proficient baseline Marines. Finally, this individual may otherwise be productive outside of the Corps, and the Corps is therefore depriving the economy of a relatively fit worker (for the individual still had to pass the entrance requirements and must therefore be generally fit, healthy, and possess a certain amount of intelligence).

Marksmanship
Marksmanship is arguably the most useful aspect of basic training. It is the only time Marines receive formal classes on shooting and marksmanship, which is absolutely necessary in modern war. Part of the reason Fox company survived and killed so many adversaries during the battle of the Chosin Reservoir in Korea was because their commanding Officer took the time to painstakingly enforce marksmanship standards before

the campaign.[1] As it stands there are about two weeks set aside to train Recruits on formal marksmanship, culminating in Table 1, and combat marksmanship, culminating in Table 2.

I shall briefly expand upon the Tables. Tables are tests on firing ranges. Table 1 requires Marines to shoot from variable distances of up to 500 yards from the sitting, kneeling, standing and prone positions. Table 2 tests Marines on combat marksmanship from near distances, firing at different body parts of an enemy (which is important if the opponent is wearing protective equipment such as a flak jacket), and reloading rapidly. These tables are intended to develop reflexes, hand-eye coordination, a killer instinct, and an opportunity to apply the fundamentals of marksmanship.

In MCT a few months later, Marines go through Tables 3, 4, 5, and 6. These tables test Marines under a facer pace with more realistic and more difficult conditions. Marines learn to walk and fire, target sudden threats quickly and accurately, buddy rush, and take out targets donned with protective equipment at close and great ranges, without knowing the exact distances (unlike Tables 1 and 2 in which Marines know exactly how far they stand from their targets). Some of these tables take place during the day, some during the night – using night vision goggles. These four tables are the culminating marksmanship events for most Marines. Infantry Marines naturally go through far more advanced exercises before deploying, but for most of the Corps, that will be it. I propose that these four additional tables be done in Boot Camp during the range after completing Table 1 and 2, freeing up additional time during MCT and increasing the combat readiness of all Marines.

---

[1] Last Stand of Fox Company, pp. 32f.

The training organization to prepare for Tables 1 and 2 at Boot Camp were a few days of class and practice all day. After the period of instruction, Recruits arrived at the range at sunrise and finished by lunch. The rest of the time during the day was more or less spent playing games, prac, and cleaning. So, half a day was essentially wasted. Further, assuming that the platoons were smaller and more motivated following my previous suggestions, the time required for administrative movements and operations, and to perform at the range would be reduced, leaving even more time.

On the flip side at MCT, each table took about a day for my company – so four days of shooting for the four tables, and about two days of classes and practice before performing the tables. On the surface it seems like a week of work – but that's because of the inherent inefficiency of the system. We woke up extremely early, put on our equipment (flak and Kevlar), and listened to a safety brief. All that would be rapid. Then a small group of individuals would go through the range. Each group went through it quickly. The problem is that there were almost 400 of us in four platoons, and only enough combat instructors to test about a dozen Marines at a time. As a result, we finished going through a range by lunch, broke for a lesson, cleaned, and prepared (quickly) the next range for that evening. We readied ourselves to go through the next range as the sun was setting and were done several hours after sunset.

It is logistically possible to displace these tables to Boot Camp without even changing the absolute number of days spent on the range. Smaller platoons would increase the pace at which the tables are performed. The afternoon could be assigned to either further classes or performing the daylight portion of the advanced tables. The beginning of the night could be dedicated to the nighttime portion of advanced tables.

Night vision goggles can be stored directly at the range to mitigate the risk that Recruits lose them. Night vision goggles are expensive; they cannot be lost. At MCT, they were assigned to each of us for about a week, so every hour we had to take accountability to ensure no one had lost theirs. The wear and tear on the equipment was enormous. It would be better for the equipment's longevity and safer if the night vision goggles were assigned before the table and stored safely immediately after.

The range's safety can be provided by the PMI's, coaches, and various other Marines who are already designated to be there. At MCT, combat instructors calmly followed behind Marines as they moved during advanced tables, ensuring that no safety violations took place and that no one moved too quickly – so that no Marine entered another's area of fire. There was about one combat instructor for every four or so Marines. At Boot Camp, our platoon had two PMI's who specialized in each Table, one coach per five or six Recruits, and a multitude of 'informal' coaches (without counting Drill Instructors). Equipped with a flak and Kevlar for personal protection, these Marines would be capable of ensuring the same safety procedures in MCT, perhaps even more effectively.

Smaller platoons would also make the end-product Marine better because each Instructor would have more time to provide individualized feedback and instruction. Preceding Table 1 and Table 2, the assigned coach judged how Recruits performed in a data-book. After shooting at their practice targets, Recruits marked where their shots had hit and the coach either adjusted their scope or gave them advice on how to improve their aim. For interested readers, the data-book can be found in this footnote.[2] In contrast at MCT, junior Marines received almost no guidance because combat instructors had too many other responsibilities to provide individual corrections. With so many individuals to get through, the pressure was focused more on efficient processing and less on improvement. By maintaining the high standards of Table 1 and 2 through the other tables, Recruits may emerge from advanced tables far more proficient than Marines currently do after the range at MCT.

Current Events

As the training matrix (Figure 6) presented at the beginning of Part 2 depicts, there is one specific event for most days; at most, there was only one event a day. For instance, on the 'gas chamber day,' my journal entry stated that we marched out to the bunker, received a briefing on using the gas mask and safety, went through the gas chamber, came back around lunchtime, and practiced drill the rest of the day. As I've made abundantly clear, smaller, more selective platoons would lead to a greater pace of learning for drill – meaning less time would have to be dedicated to it to reach an equivalent level of expertise. As a result, smaller platoons mean more than one event could be fit into a single day. The afternoon of the gas chamber, we could have also gone through at least one assault course and a hike.

Each event in basic training exists for a reason, I propose that we go through them at a more rapid pace, leaving room for other events in the latter stages of training. At OCS I vividly remember that the training

---

[2] U.S. Marine Corps, "ANNUAL RIFLE TRAINING DATABOOKM16A4 SERVICE RIFLE/M4 CARBINE WITH RIFLE COMBAT OPTIC (RCO)AND BACK-UP IRON SIGHT (BUIS), NAVMC 11660 (Rev. 02-12)," https://www.trngcmd.marines.mil/Portals/207/Docs/wtbn/ART%20NAV MC%2011660%20REV%2002-12.pdf.

tempo was much faster. For example, one day we awoke in the dark, having camped a few miles away from the barracks. We hiked back to the main compound in time to do a leadership reaction course in the morning. We ate whatever remained of our MRE's after the course, then hiked into the woods to do a fire team assault course, which we finished in the mid-afternoon. I don't remember what we did after that – likely drilled and classes – but the point stands that we underwent a hike and two courses in about half a day! With higher standards, the Enlisted side could follow a similar tempo. A few classes, MCMAP, a bayonet assault course, and PT could all be done in one day in Phase I, rather than over the course of four.

Small Unit Tactics
The end of MCT was marked by an event called "live fire and movement," which involved buddy rushing forward in a designated space and firing to cover the movement of your 'buddy.' It's certainly an important skill in modern war, for buddy rushing can be employed to protect one another, keep the enemy pinned, attempt to achieve fire superiority, and to maneuver to a more advantageous location. Yet it is also possibly the most basic technique. Further, Marines are not trained to use buddy rushing to achieve anything, they're simply taught to fire when their buddy runs forward. After four months of training, Marines are not prepared in any way for the modern battlefield if the culminating event consists solely of two Marines running between predetermined sandbags and shooting straight ahead.

In Boot Camp, the extent of small unit tactics were 'fire team assault courses,' which we did a few of. These also consisted of clearly established courses and obstacles at which the fire team stopped, calmly lined up, and moved past the next obstacle. Each step was formulaic and predetermined. About the only requirement in those courses was the stamina to war cry for a few minutes as the team moved forward. Basic Enlisted courses were centralized, ordered, completely controlled, and required no thought whatsoever. Such a setup is necessary as a first step but should be followed by more challenging real-life scenarios. As it stands, the entirety of classes are the exact opposite of "an environment of chaos, uncertainty, constant change, and friction."[3] As I have discussed, it is possible that these Marines will be sent into combat with little to no additional training; knowing how to buddy rush on a controlled range is simply not enough to survive or thrive as a rifleman.

In contrast, by the end of PLC I at OCS, Officer candidates were going through complex courses in fire teams over variable distances and learning to use fire teams to overcome various problems thanks to the leadership reaction course and rotational leadership requirements. While junior Marines do not need to be on the same tactical level as their Officers, it is ridiculous that in *one month* of training, Officers are more advanced in small unit tactics (that they will rarely take part in – since a Lieutenant should theoretically be able to serve as Platoon Commander) than their Enlisted counterparts after *four months*.

I propose that the fundamentals of buddy rushing be taught in Boot Camp alongside fire team tactics so that MCT may be reformatted to teach advanced fire team and squad sized operations. In practice, that should not be so difficult to implement in Boot Camp, for we were taught the organization of a fire team during Phase II along with land navigation, we just never used or applied the fire team in any meaningful manner. Buddy

---

[3] U.S. Marines, *MCDP1: Warfighting* (North Charleston, SC: Create Space Independent Publishing Platform, 2010), pp. 53.

rushing is simple to teach and could be implemented relatively early on without live ammunition to mitigate the risk of injury.

Once buddy rushing and simple fire team tactics have been taught, they should be tested in environments that require quick reaction and thoughtful decision processes from the fire team leader. To accomplish this objective, I propose four events:

- A fire team assault course, similar to that currently staged at OCS, adapted for local conditions.
- A fire team operation on a tract of unorganized terrain.
- A fire team operation combining land navigation and buddy rushing.
- A Leadership Reaction Course similar to the one at OCS.

The fire team assault course tests endurance whilst presenting an opportunity to practice small unit tactics and utilize basic cognitive reasoning to move over terrain and overcome obstacles. It's a good way for Recruits to apply the various fire team formations, assault an enemy position, implement simple security protocols, negotiate diverse obstacles in various manners, and receive feedback on their performance. It also allows Recruits to practice working as a fluid team and to apply small unit leadership under stress.

A fire team operation on a tract of unordered terrain took place in the woods of Quantico at OCS; we set off over streams and hills, deployed in different formations based on the threat level and natural obstacles, and practiced a simple assault of a location. Where the previously mentioned Boot Camp assault courses were organized with clear cut obstacles, this event took place in a more natural environment, leaving room for flexibility of thought and operations. Such a course could also be staged in urban as well as rural terrain, for a city provides tactical difficulties not found in natural environments. If a course is implemented in an urban terrain before MCT, a lesson would also be required to teach Recruits to clear houses and develop a conception of the additional dangers and difficulties of fighting in a built-up location.

Combining land navigation and fire team offensive tactics would exercise the concepts of land navigation in a more realistic scenario. Recruits would have to locate their objective given coordinates on a military map, decipher and reach that objective, then capture it as if it were occupied by hostiles. To do so, the fire team must be thoughtful – for instance it should not directly attack an uphill prepared position in a column formation. Rather, they must change their disposition in accordance to their environment and needs. For instance, the team could march forward quickly in column. When the team leader believes the team near the objective, the team deploys in a wedge, identifies the best approach, and finally assaults the objective in a skirmisher's formation while carefully maneuvering across the natural environment and buddy rushing. Thus, not only will all three major fire team formations have been practiced, but the relationship between land navigation and combat would have been bridged.

Finally, a leadership reaction course (LRC) is sorely needed for Enlisted personnel. The LRC is an obstacle challenge – with no clear solution – for a fire team. The fire team leader is given a situation to go along with the obstacle by a monitor. After receiving the information, the leader provides the relevant information to the team along with a rough plan to overcome the obstacle, thus testing them on their ability to issue a five paragraph order.[4] Most plans fail, the goal is for the leader to continue coming up with new plans, to maintain the team's morale, and to do so with confidence and decisiveness. Different types of leadership techniques can be tried – from active micromanagement, to democratic, to laissez faire – with varying results. There was a version of the LRC in Boot Camp during the crucible, but there was no five-paragraph order, no pressure, no focus on leadership, and no feedback. It was an LRC only in name, not in content. The entire 'stick,' or squad I was with went through a single obstacle once, which we did not come close to accomplishing. The real purpose of the event was for the Drill Instructors to send Recruits to 'resupply,' to task Recruits to run with ammo cans up and down a road. While it tested one's endurance, this version of the LRC had absolutely nothing to do with problem solving or practicing leadership. Recruits should go through LRC's *at least* once as leaders to implement small unit leadership, giving five-paragraph orders (which we never did either in Boot Camp or MCT, but could have learned easily), decision-making under pressure, and creative thinking. Recruits should go through many LRC's with variable obstacles several times to improve and try new strategies rather than simply be tested on their skills.

Logistically, the first and last courses would require the most work and investment to construct the necessary obstacles. A fire team assault course could be established by making use of a few crucible obstacles with newly constructed scenarios. The other two could take place during what was Phase II for me, on the West

---

[4] A five-paragraph order is the most common order given to subordinates. The order converts the leader's plan into action, provides orientation to the unit, gives direction to the efforts of their unit, and provides specific instructions to subordinate elements. It's a relatively long and arduous order to give, consisting of multiple different parts, meaning giving and receiving the order effectively takes practice.

coast. The rolling hills of Camp Pendleton and the swampy woodlands of Parris Island would be perfect and require little modification. For the LRC, a few well briefed Lance Corporals and Field Instructors could preside; I set up an LRC during my MOS school, which showed me that Corporals and junior Marines were more than capable of moderating the obstacles professionally. The various courses would require one Marine for each Recruit fire team to grade and provide feedback on performance. If Recruits undergo an LRC several times, by the latter iterations they might be able to critique, or even grade one another.

Rotational Leadership
Leadership at Boot Camp should be rotational in order to let all Marines improve their abilities, be judged by their effectiveness at leading, and increase respect for future leaders by enabling Recruits to better understand the oftentimes difficult decisions leaders must make. Additionally, leadership opportunities that are substantive and impactful should also be presented so that leading actually matters and leadership skills are cultivated.

Leadership at Boot Camp was static. Within the first and second week, the leadership of my platoon had more or less crystalized, the same was true for the other platoons in the company. The individuals that were put in a position of authority were the 'model Recruits:' good at drill, physically fit, and motivated. My Senior Drill Instructor told me that Drill Instructors wanted leaders for other Recruits to look up to, to emulate. In my experience, inefficient, ineffective Squad Leaders and Guides (the Recruit Leadership) greatly affected the efficaciousness and tempo of the platoon. Beyond leading drill (being at the front of the platoon) and routines, the Recruit Leadership was responsible for enhancing the abilities of Drill Instructors to move other Recruits quickly and maintain discipline. They were also held responsible for mistakes made by other Recruits. The Recruit Leadership was often IT'd or reprimanded for the failures of their 'subordinates,' even if there was no connection or relationship.[5] The only time other Recruits exercised any leadership billet was as 'stick leaders' during the crucible, usually for a couple hours. By and large however, their job was inconsequential and thoughtless. For instance, leaders ensured that we were all taking a knee on the same knee and kept the formation tight as we hiked.

The current system of Recruit Leadership is based on having the most disciplined platoon possible, rather than improving individual Marines. But the Marine Corps does not fight using first-generation tactics, and the platoon does not stay together after Boot Camp. Building a discipline platoon is therefore less important than improving individual Marines. Smaller platoons would increase the efficiency and speed of the platoon, a shift should be made so that each Marine improves their leadership skills. Fortunately, a well-established precedent already exists at OCS, rotational leadership. Candidates are placed in charge of leading their peers in simple tasks such as PT and maintaining accountability. By the end of PLC I, candidates had learned to occupy most of the simplified billets of a platoon quite effectively – from candidate fire team leader all the way to candidate platoon commander - and

---

[5] Leadership at MCT was far worse than at boot camp, it was almost wholly nonexistent; the guide and squad leaders' billets existed to pass out food and try to maintain some semblance of discipline and silence when the platoon was on standby for unending durations, they were picked pretty much at random by the instructors; I remember for example one of my instructors asking a Private his name and saying 'oh that's a good name, you're a squad leader.'

only for a set period of time to give everyone a chance to exercise military leadership.

While it is neither optimal nor necessary for Recruits to do the same, it would be beneficial if the billets of Squad Leaders and Guide were to rotate every two or three days (except approaching Initial and Final Drill), and for additional billets such as Recruit platoon sergeant to be added. Not only would Recruits have the chance to improve their leadership and practice accountability, they would come to better understand the procedures of military communication and command. Recruits would be forced to exit the general population of the platoon and lead their peers, thus experiencing something that is far out of their comfort zones. The long-term efficacy of the platoon would likely be improved since each Recruit might understand the difficulties faced by billet holders attempting to fulfill their duties. As it stands, most Recruits tend to dislike their semi-permanent Recruit leadership, and respond out of fear of the Drill Instructors, rather than motivation to fulfill the mission. If that leadership were to constantly change, the division between and aversion for Recruit Leadership would not emerge.

The theory behind Recruit leadership must be revamped. Squad Leaders and Guides provide immaterial leadership. Removing them entirely would in fact not change much about a platoon. After over a month in Boot Camp, most Recruits understood the basic tenants of training. By the end of Boot Camp however, the duties of Recruit Leadership remained essentially the same. By and large the most important duties were to drill well and be the most physically fit Recruits. Surely by Phase II or III, additional day-to-day duties such as leading in PT, assuring accountability, and training fellow Recruits in drill could be added. In these Chapters, I also suggest the addition of a few key leadership events, notably the LRC. Recruits should be graded on their performance during these events and seek to improve their leadership capability and style. In sum, I believe practicing leadership in specific instances is necessary in Boot Camp.

I realize I am importing many elements from OCS. Let me be clear – the goal of Boot Camp is not the same as OCS. They are supposed to be fundamentally different: Boot Camp builds basic Marines, OCS tests whether a Candidate can become an Officer of Marines. Officers are evaluated on their leadership potential by putting them in tough conditions where they may demonstrate leadership capabilities. I find some of these conditions to be particularly lacking in basic training because, as I've been arguing, leadership training and opportunities to exercise it are missing and are necessary. The reason I source my suggestions from OCS is that the Marine Corps has already decided that these events and procedures matter and is highly proficient at implementing them. Because of existing practices, executing my suggestions would be cost effective and leverage existing expertise.

Chapter 6: Training Changes in Marine Combat Training

I did not cover MCT – Marine Combat Training – in the Boot Camp portion of my book. I did not take careful notes of MCT while there but remembered most details and events quite well. Instead of creating a day-by-day journal like I did for Boot Camp, I shall discuss the purpose of MCT, a brief synopsis of what MCT was like for me, and a critique of the training program. The program will be contrasted with the suggestions made in various sources from the commandant's reading list to suggest an alternative training program that is more in line with the experience and advice of successful commanders and Marine Corps doctrine, in order to achieve the necessary skills to operate as a rifleman on the battlefield and successfully conduct maneuver warfare.

According to the official website of the Marine Corps, the mission of MCT is to generate riflemen who "possess a foundational understanding of, and their role in applying, the Marine Corps' Warfighting ethos, core values, basic tenets of maneuver warfare, leadership responsibilities, mental, moral, and physical resiliency in order to contribute to the successful accomplishment of their unit's mission."[1]

---

[1] https://www.trngcmd.marines.mil/Units/West/School-of-Infantry-West/Marine-Combat-Training-Battalion/; mirrored in the synopsis shown for MCT – East: https://www.trngcmd.marines.mil/Units/South-

To achieve this goal, over the course of a month, we went through four shooting tables, a few hikes, learned how to clear a house, buddy rush, how to operate an M240B, how to throw grenades, and the basic elements of patrolling and defense. We also received a class on "combat hunter," and a class on nutrition. On its surface, this seems like quite a lot accomplished – but it really was not. The vast majority of our time in MCT was wasted moving vast platoons of 100 Marines and 'standing by.' The most vivid image I have of MCT was sitting in the squad bay, reading *Promise and Power: The Life and Times of Robert McNamara*, waiting to be told to go outside, go eat, or to go to sleep.

A description along with my opinion of the events is as follows. The 5, 10, and 15-kilometer hikes with 70-pound packs were useful to build our stamina and to reach the shooting range with the necessary equipment. Once we reached the range, we set up and went through Tables 3-6. I've discussed these tables in the last chapter. MOUT – military operations in urban terrain - was one of the most practical skills we learned, but we could have used a lot more time on this. We spent one day learning to 'pie' a window and enter a room, and another day 'attacking' an artificial town. The attack took place without any tactical considerations. Packed fire teams ran in a straight line to the town while a 'squad' of 30 Marines 'provided cover' from a nearby hill in full view of the hypothetical enemy. A couple riflemen in the town or on another hill, or a few well-aimed mortar rounds would have wiped us all out. There was no plan generated by any junior Marine. The fire and movement training – buddy rushing – was a building block, but it represented the height of our tactical knowledge by the end of MCT. Learning how to operate

Atlantic/SOI-E/Units/Marine-Combat-Training-Battalion/, Accessed June 2, 2018.

the M240B was useful and interesting. Learning how to pull the pin on a grenade and the proper procedures for throwing it using blue bodies in lieu of real grenades was also necessary. We also learned to patrol and identify signs of IED's - a class we received twice in Boot Camp – a requisite skill, especially because of U.S. engagements in the Middle East. Finally, we learned and practiced digging fox holes (called fighting holes in Marine manuals) and the basic theoretical tenets of Marine defensive warfare.

Weekends at MCT were pathetic. We received a couple hours of liberty (time off). That liberty consisted of standing in line to get food, then standing in line to get a haircut, then standing in line to buy necessities at a tiny PX, before running back to the barracks in time to check in. Liberty was really a waste of time, a substitute for the weekly Boot Camp 'PX call and haircuts.'

The problem with MCT was not the content, which as I've indicated, was practically useful, but rather how much more useful MCT could be - especially compared to how thoughtless and inefficient it currently is. As I have discussed previously, the current output Marine is not trained for modern combat; buddy rushing is their sole available "tactic." When contrasted to the prescriptions of MCDP1, MCT just does not do what it's supposed to do.

There were several reasons why MCT was a failure. First, the platoons were massive. It took forever to move the company from one location to the next, not to mention the time required for actual events; the unwieldy size of units severely restricted the number of events the company could go through. Second, control and order were prized above all else. On the one hand, this is to be expected – junior Marines are supposed be taught specific basic skills and the command wants to ensure the environment is safe and that those basic skills are being taught and taught correctly. This became an issue when considering the third major cause I identify: over-centralization. It would have been fine if decisions had been delegated to the Platoon Sergeants – but instead they were all made at the company level or higher. No detail was managed at the local level, there was a strict set of events which the company had to attend at specific times, and no one could act without orders from above. As a result of the combination of platoon sizes, obsession with control, and over-centralization, we spent hours moving the platoon between locations, immeasurable time for accountability after every movement and event, reorganizing the platoon or company again and again, and rushing Marines through a few simple lessons. After one month of hard work, Marines could be relatively proficient at small unit tactics and combat – instead they had only learned to pie windows and throw blue bodies![2]

Marine Corps doctrinal publications and various legendary commanders discuss recommended ways to train men to gain experience and prepare for war. In this section, I showcase the recommendations set forth by *MCDP1* and *Marine Corps Tactics* before introducing the recommendations made by Adolf Von Schell, Erwin Rommel, and Richard Hooker – all sources from the Commandant's Reading List. I support my proposal with additional evidence from various other sources from that same reading list. Based on these works, I introduce my proposal for a reformed MCT. I take as a given the arguments made in the last chapters and build off that established context: that maneuver warfare requires intelligent subordinates to take initiative and that combat continually becomes more decentralized over time –

---

[2] To 'pie a window' is to use the sights on one's rifle to 'slice' a window, thereby taking in a picture of the room behind and reducing the probability that a surprise may emerge from a hidden corner. One pies a window or a hatch opening as one is passing in front of it.

leading to depreciating command and control capabilities.

MCDP1 states "In order to develop initiative among junior leaders, the conduct of training – like combat – should be decentralized... As a rule, [senior commanders] should refrain from dictating how the training will be accomplished... exercises should approximate the conditions of war as much as possible; that is, they should introduce friction in the form of uncertainty, stress, disorder, and opposing wills. This last characteristic is most important; only in opposed, free-play exercises can we practice the art of war."[3]

The sentiment is echoed by *Marine Corps Tactics*. "Any approach should be adaptable to all echelons and to all grades. The environment should be one that is challenging and conducive to creative thinking. Like all preparation for war, training should reflect the rigors of that environment."[4] Later, it says, "Exercises should provide realism. The means to achieve tactical realism are competitive free-play or force-on-force exercises. Whenever possible, unit training should be conducted in a free-play scenario."[5]

Von Schell supports these proposals, for "there is no situation that our imagination can conjure up which even remotely approaches reality. In peace we have only grammar school tactics. But let us never forget that war is far more advanced than a high school. Therefore, if you would train for the realities of war, take your men into unknown terrain, at night, without maps and give them difficult situations." [6] Such a proposal is directly

---

[3] U.S. Marines, *MCDP1: Warfighting* (North Charleston, SC: Create Space Independent Publishing Platform, 2010), pp. 39.
[4] U.S. Marine Corps, *Marine Corps Tactics* (New York: Cosimo, 2007), pp. 119f
[5] U.S. Marine Corps, *Marine Corps Tactics,* pp. 125.
[6] Adolf Von Schell, *Battle Leadership*, (Battleboro, VT: Echo Point Books, 1933), pp. 63.

paralleled by Rommel: "Train in time of peace to maintain direction at night with the aid of a luminous dial compass. Train in difficult, trackless, wooded terrain. War makes extremely heavy demands on the soldier's strength and nerves. For this reason, make heavy demands on your men in peacetime exercises."[7] Likewise, Hooker writes: "Those small units who do the fighting, killing, and dying must be capable of shouldering the greater demands of maneuver warfare. They must think and act faster. They must approach infiltration and penetration into rear areas, night operations, decentralization, and the increased friction and uncertainty of the nonlinear battlefield with confidence and resolve."[8]

In addition to free play exercises, units should seek continuous professional military education. "Tactical success evolves from the synthesis of training and education – the creative application of technical skills based on sound judgment,"[9] education is imperative. *Marine Corps Tactics* suggests historical studies which "provide the most readily available source of indirect experience in our profession. These studies describe the leadership considerations, the horrors of war, the sacrifices endured, the commitment involved, the resources required, and much more…. Group discussions help to expand the insights into leadership and battle that we have gained through individual study."[10] Von Schell concurs, "The men, although well trained and of high morale, were inexperienced in war and reacted strongly to early impressions. The reason may well be that they had not been psychologically

---

[7] Rommel, *Attacks* (Provo, UT: Athena Press, 1979), pp. 8.
[8] Richard Hooker Jr., 'Implementing Maneuver Warfare,' in Richard Hooker Jr. *Maneuver Warfare: An Anthology* (Novato, CA: Presidio Publishing, 1993), pp. 230f
[9] U.S. Marine Corps, *Marine Corps Tactics*, pp. 121
[10] U.S. Marine Corps, pp. 120

prepared for the severe trials they were called upon to undergo.

The conclusions to be drawn from this is obvious: we must teach our men in peace that battles differ greatly from maneuvers and that there will often be critical periods when everything seems to be going wrong. It is exceedingly difficult to teach men what to expect in war, but something along this line may be accomplished if we study military history and teach its lessons to our soldiers."[11]

Von Schell goes on to say there is a less formal and directed manner to educate the men, by embedding experienced men with greener units. "The German troops were young and had only undergone a short period of training. They had... but three months of training behind them. However, they were intermingled with men who had already had some war experience, and who at least knew those first impressions that war brings. These veterans regarded themselves as instructors to their young comrades; they felt a certain responsibility for them. Because of this feeling the value of the old soldiers was markedly increased, while the inexperienced men developed rapidly under their instruction. Although only one-fourth of the men were experienced, their influence was sufficient to give the entire organization a veteran character."[12]

We can sum up these suggestions as the following. 1. Practice how you fight, in free-play exercises under difficult conditions. 2. Educate the men using military history so that they may draw upon the lessons of the past. 3. Use veterans to teach troops informally so that they impart their experiences to them.

---

[11] Adolf Von Schell, *Battle Leadership*, (Battleboro, VT: Echo Point Books, 1933), pp. 39.
[12] Von Schell, *Battle Leadership*, pp. 46.

Based upon this information, I propose that MCT be reformed away from its inefficient centralized current state to a massive informal free play month long exercise. Instead of giant 100-man platoons, platoons should be split into squads and assigned to a junior NCO – a Corporal or Sergeant. For the duration of combat training, this squad will travel across a large tract of terrain, engage in combat with other squads, and accomplish objectives; all the while receiving criticism, instruction, and guidance from their NCO. After being issued their equipment, squads will split apart, and only reconvene into platoons to resupply periodically, to take final tests that ensure the NCO taught their unit properly, and for graduation. With Tables 3-6 shunted to Boot Camp (see Chapter 5) there is no longer a need for a range week and the associated required safety conditions.

This is a radical proposal considering how centralized MCT is now and how much control would have to be relinquished by senior NCOs and commanders. Part of the prerequisite for this proposal to be accomplished is that the NCOs in charge of the squad will be held responsible for the actions of their squad, and not the Officers or NCOs of the platoon or company since the squads cannot and should not be under the constant supervision of any command.

To be an instructor at MCT currently, NCOs must attend a rigorous course called Combat Instructor School. In this reformed MCT, NCOs would not go through Combat Instructor School, but they would be extracted directly from infantry units stationed in nearby bases. On the West coast: Twenty-Nine Palms and Camp Pendleton. On the East coast: Camp LeJeune. It would give them a chance to impart their knowledge to inexperienced Marines as Von Schell suggests. It would allow them to teach and supervise a unit with a great degree of autonomy for one month, improving their leadership abilities. NCOs would be rotated quickly away so that they only spend the one-month away from their units, which reduces the impact upon the unit and spreads the benefits of command to many NCOs.

I envision MCT as a radically decentralized training period during which new junior Marines are guided to constantly exercise their leadership abilities and learn through both informal training from their NCO and from practical applications of tactics. Overall the goal is for junior Marines to learn and exercise their own judgment, to apply lessons and ideas as they see fit, and to learn from their successes and failures. In this vision, MCT could be just as long as it currently is – or shorter. Its length could easily be cut by one week by removing training Week 4. The following section is a detailed plan of my proposal.

––––––

Week 1

Tuesday to Friday
The process for the beginning of the first week would be similar to current MCT. After checking in, entrants will be processed administratively and medically. Marines will then go to supply to receive their gear, and the armory to receive their weapons – which should include old M240's, preferably one or two per squad (depending on current availability). Finally, Marines will be issued MRE's, maps, compasses, and map pens. Marines will be split into squads and introduced to their NCOs, then briefed on the rest of MCT. In total, the processing period should take no more than a few days.

If possible, classes would take place immediately. The classes taught during this span of time should be combat hunter, a refresher on land navigation, patrolling, and IED identification. Time allowing, lower priority classes could be conducted – such as 'force fitness' and nutrition – which are repeats of courses given in Boot Camp. It is important that Marines receive enough sleep during the night so that they may engage with and comprehend the material.

———

Saturday and Sunday
During this first weekend, Marines will be given time to arrange their affairs, purchase any needed items from the MCX, speak to their loved ones, and pack their gear for the training ahead.

———

Week 2

Monday
In the morning, with all their gear packed, each squad will depart from the barracks on their own. For the duration of their time in the field, Marines will carry their gear in main packs or assault packs. While the squad is supervised by an NCO, the actual Squad Leader - the one who makes decisions - will be a junior Marine. Every two days, leadership will rotate for both the squad and fire team leaders. The squad will be provided five-paragraph orders at regular intervals by the NCO, with special emphasis on the commander's intent. For central command to retain some control, these orders can be issued to the NCO over the radio, by phone, or be prewritten and unveiled at regular intervals. These orders should be realistic, paralleling similar orders issued during actual combat, including the possible presence of hostiles and a context for the movement.

The first hike will be largely administrative to reach a suitable area to camp. Each squad will camp in a different location. Squads will not intermingle. Crucially it is not the NCO who leads the squad to its destination, but the Squad Leader based upon their map. Land navigation will thus be applied to more than just finding an ammo can. Once they have arrived, the squad will set up camp and receive lessons from their NCO. This is the most formal part of lessons. Lessons will cover what was prescribed but may not have been covered in Week 1. Once these initial lessons are completed, the NCOs will cover the basic tenets of defense, field hygiene, and how to operate the M240B. If necessary, the NCO will be provided materials with which to conduct this teaching (i.e. a lesson plan). The first day will conclude after this period of instruction is over, allowing the Marines to rest from the exertion of the first hike.

Tuesday

The first full day in the field will consist of lessons and practical applications. The NCO will ensure that material is understood, then supervise the squad to confirm understanding of the previous day's and today's lessons. The latter consists of the principles of patrolling, a refresher on buddy rushing, defense, fire team tactics, squad tactics, and how to handle a grenade (using blue bodies or stones).

The NCO can choose to split the squad into separate units. A fire team can be tasked out to practice a perimeter patrol while another digs a skirmisher trench and prepares to be quizzed on the premises of defense, while a third practices operating the M240B. During periods of rest, Marines should be either reading or being taught by their NCO. Beyond the standard knowledge handbook that is always provided when Marines arrive on base, Marines should also read *MCDP1: Warfighting, The Defense of Duffer's Drift* (which highlights principles of small unit defensive tactics and is on the Commandant's Reading List), and *Marine Corps Tactics*. It only took me a few hours to get through all three books, which are foundational resources for Marine Corps theory of small unit combat operations. Marines should either be issued these works, have significant excerpts of these cited in their 'knowledge,' or be made to purchase these upon arrival to MCT. While eating and resting, the NCO should ensure that Marines are being productive with their time.

Given a total of sixteen hours in the day, minus two for chow, under intensive study each squad should be able to understand the basics of small unit operations. The pace of training should be intense. Not a moment should go to waste, a fast-paced operating tempo must be maintained since there is so little time in MCT. "Standing by" should never occur.

———

Wednesday to Saturday

Marines will begin implementing the lessons they have learned by performing brief structured exercises developed by the squad NCO. In accordance with the prescriptions for small unit training set out by Lieutenant General Arthur Collins Jr., training problems will be kept short and concise; revolving around specific terrain features. While General Collins states that small units can perform two or three such problems a day, he also says that training should not be rushed to allow for substantial critique, coordination, and planning.[13]

Each squad will receive orders from their NCO which mirrors – to the greatest extent possible – real combat orders. Squads will be tasked to attack hills, hypothetical enemy camps, towns, convoys, defend

---

[13] Arthur S. Collins Jr., *Common Sense Training: A Working Philosophy for Leaders* (Novato, CA: Presidio Press, 1978), pp. 134f.

positions, withdraw, evacuate, and various other possible maneuvers. It is crucial that the NCO not lead the squad, but that the squad lead itself, receiving criticism and feedback after an operation is completed. The squad itself should choose every possible aspect of its operations. a stealth-based mission, for instance, could require the shedding of personal protective equipment. Mistakes should be made, adaptation should occur, and successes noted.

Two to three such events should take place a day. During periods of rest, as always, Marines should be studying, practicing the M240B cycle of operations, cleaning their weapons, and learning what they can from their NCO. The camp should change locations regularly to practice packing quickly, policing an area, hiking, and land navigation. Everywhere the squad goes they should hike with their assault packs or main packs to sustain and improve the stamina developed in Boot Camp. Units in modern war can be on the move for enormous periods of time across difficult terrain, often with little sleep. As Rommel writes, "in the retreat of September 13, a march of twenty-seven miles was planned for troops who had been on outpost duty the previous night. The many halts and the assistance-required by the bogged-down trains and artillery made this movement all the more difficult. The battalion was continually on the go for more than twenty-four hours."[14] The cumulative distance travelled over the course of these free play exercises should surpass the current 5, 10, and 15-kilometer hikes currently performed in MCT, eliminating the need for those hikes as formal events.

––––––

Sunday

––––––

[14] Rommel, *Attacks* (Provo, UT: Athena Press, 1979), pp. 49

As the designated rest day, this will be the first day squads will intermingle again when they meet for optional religious services. At MCT, I recall we stopped on Sunday in a set of bleachers out of the base on a road to the range to hold religious services. I don't see why it wouldn't be possible to establish such a central location for all squads to convene. This central location can also be a resupply day where MRE's and mail are dropped off. It would also be an ideal time to ensure the NCOs are doing their jobs by providing an opportunity for Marines to report poor behavior and by spot checking – asking a sample of Marines in each squad questions to that end. Once services are concluded, squads should return to their areas to clean their weapons and study.

———

Week 3

Monday through Friday
Since two to three exercises had been performed every day, Marines have now performed between eight and twelve small unit tactical exercises as a squad. If the NCO feels their squad needs more practice, they should continue these exercises for another day. However, if the NCO believes their squad has developed basic proficiency, an additional level of realism should be added for this week by setting squads against one another. Instead of attacking empty positions, those positions will be defended by another squad. As units move from one location to another, they can be ambushed. At night, squads become vulnerable if their fire watch does not notice an enemy sneaking up on them. NCOs should guide their squads subtly to generate clashes. Now, squads that don't change resting locations

or get sloppy and let their guard down become vulnerable to destruction.

These free play exercises are the most important aspect of training. This entire week should be spent on heightened alert. If a conflict is in progress, it should not be ended but allowed to continue until fully resolved. This week should involve the implementation of all that was previously practiced, squads should move about in patrolling formations, use simple squad tactics to perform operations, move with stealth when necessary and buddy rush when assaulting positions 'in combat.' At night fire watch must be active, but NCOs can also rouse their squads and have them practice nighttime maneuvers. If possible, squads should fortify and defend positions, while other squads attempt to breach such positions during day or night. These exercises must be free play. Once mission intent has been established, the exercise must flow. Allowing Marines to choose their positions and to try to make the most of themselves as they would on the battlefield. As Rommel wrote: "the Rommel detachment, by taking advantage of the smallest irregularity of the terrain, succeeded in capturing and defending the crest of the heights eleven hundred yards behind the hostile front.[15] It is far easier for individuals to find crooks and crannies, and take out enemy units when they are capable of doing so, than if they were a part of a large massive unit moving as if it were a single organism. Squads must learn that different tactics and approaches should be developed based on available resources, terrain, and context.[16] Group circles to discuss operations should occur as individuals reflect on what worked and did not work.

───────

[15] Rommel, *Attacks*, pp. 147.
[16] U.S. Marine Corps, *Marine Corps Tactics* (New York: Cosimo, 2007), pp. 150.

Saturday and Sunday
On Saturday, the commander can choose to either: 1. continue exercises 2. go back to the base and have an off day where Marines receive haircuts, telephone their families, eat and relax – the typical Saturday currently enjoyed by Marines – or 3. stay in the field to rest, study, and clean. Sunday would be the same as the week before, featuring the opportunity to attend religious services and a resupply for food, equipment, and ammunition.

———

Week 4

Monday through Friday
MCT is currently a full month long. This final week of training can be added or removed. If it is implemented, it would be a continuation of the free play exercises of week three. Were it removed to cut down on a week of training because Marines are urgently needed, we would simply stop week three after the events on Thursday and jump to the Friday discussed below.

———

Friday
On Friday of week four, squads will hike back to the barracks at the School of Infantry and provide their Marines the first chance in nearly two weeks to shower, shave properly, and sleep comfortably. In the afternoon, squads should finish reading their books and study for the final exam.

———

Saturday

The final examination should take place in the morning consisting of three parts: 1. Making sure M240B operations are known, 2. A multiple choice, written test, on material scheduled to have been covered by NCOs, 3. Basic fire team and squad tactics. Once the exam is completed, the M240's will be cleaned once more and returned to the armory. If there's still time, Marines will be given liberty to socialize, eat, and get haircuts.

———

Sunday
In the morning Marines will clean their rifles before turning them into the armory. If possible, they will also receive their orders today (for where their MOS school is and how to get there), then be given time off to go to religious services and receive haircuts.

———

Week 5

Monday
Marines will go to chow, practice for graduation, and orders will be given to those who have not yet received their orders, then all gear will be returned to supply.

———

Tuesday
Final administrative issues will be dealt with, then graduation.

Tentative Training Matrix

	Monday	Tuesday	Wednesday	Thursday	Friday	Saturday	Sunday
Week 1	N/A	Arrival and processing				Final individual preparation	
Week 2	Depart from SOI + first lessons	Lessons and practical applications	Event-based lesson implementation and small unit tactic practice				Services + resupply
Week 3	Free play exercises					Rest + services + resupply	
Week 4	Free play exercises				Return to SOI + rest	Final examination + return M240B	Return rifles + receive orders
Week 5	Graduation practice + return gear	Graduation	N/A				

Logistical Issues

Though I propose that NCOs not have to go through Combat Instructor school, it would make sense for them to go through a few basic courses to ensure their squad is in capable hands – for instance a first aid course to know how to deal with minor wounds and accidents; and a class on how to operate radios and flare guns, for NCOs should be equipped with radios and back up batteries to communicate with each other and with their command. Should the radios fail during an emergency, they will also have a flare gun. Several urgent care vehicles should be made available to bring injured Marines to medical or the hospital.

Marines will not be equipped with live ammunition. Instead, they will carry full combat loads of blanks or, if financially feasible, either simulation rounds or MILES (multiple integrated laser engagement systems). The best possible outcome would be to equip Marines with simulation rounds in week three of MCT in order for the decentralized free play exercises to be as realistic as possible when squads clash.

Water will be available at various water bulls. However, since squads will be in competition, an ambush at a water bull may be likely, unless command decides to make it a conflict-free zone. Marines will otherwise be equipped with water-purification tablets to use at rivers if the terrain boasts natural sources of water should they wish to opt out of a conflict. The water bulls will be scattered across the map at disclosed locations and should be checked and refilled regularly.

Supervision of the squads by higher echelons of command is inherently next to impossible in this scenario. NCOs will have to be trusted to guide and supervise their squad without being checked upon from above during the process. The supervision must instead be made before and after the squad engages in exercises. Junior Marines should be told what to expect from their NCO so that they may hold them accountable. At the end of the exercise, Marines will testify whether their NCO has behaved correctly. During the Sunday of week two and three, Marines will also have an opportunity to report any inappropriate or subpar behavior. The final examination on Saturday serves to guide the NCOs as they teach a basic set of information to their squads, as well as a check of their abilities to have done so. Though I have allocated only one NCO per squad, perhaps experience will reveal three to four weeks of constant leadership from a single individual is overly demanding, or that two NCOs are necessary to keep each other in check.

By and large however, the radical decentralization and lack of supervision, while rife with potential repercussions, is in fact necessary for the development of maneuver warfare. The lack of a specific curriculum is intentional. I only suggested reading major resources and not detailed material therein. The intent is that NCOs will impart their knowledge and keep on challenging their squads as they see fit and as circumstances dictate – not as a prewritten set of orders state. General understanding of topics is the goal, not the memorization of moot details that are quickly forgotten after MCT is over. Short term memorization of topics does not make Marines better fighters; improving their leadership, judgment, and understanding of their job as riflemen does. As retired Marine Colonel Michael Wyly wrote, "Teaching maneuver warfare is like doing maneuver warfare. When the teacher begins, he does not know how far he will go in an hour or how many turns he will take in how many different directions. His mission, after all, is to teach students to think, to exercise judgment. It is not to teach a repertoire of attacks or formulistic procedures... The teacher must proceed at whatever pace will keep his students challenged and exercising their minds. The course is not about imparting knowledge. It is about teaching judgment."[17]

Honor Squad

Since units do not operate as platoons, it would make sense to replace the honor platoon designation with that of 'honor squad.' The method to determine the best

---

[17] Michael Duncan Wyly, 'Teaching Maneuver Warfare,' in Richard Hooker, ed., *Maneuver Warfare, an Anthology* (Novato, CA: Presidio Press, 1993), pp. 250.

squad will be both *a.* quantitative and *b.* qualitative. The weighting would not be equal, in order to incentivize squads to perform as many actions as possible during the free play exercises to receive more points. The quantitative aspect will be based upon the test scores from the written test, the final examination. Perhaps points can also be awarded for rifle cleanliness upon being passed into the armory to incentivize Marines to clean properly. The maximum number of points possible is 15.

Requiring far more finesse is the qualitative aspect. Because squads will be engaged in mock combat against one another, they must exercise physical and intellectual prowess to succeed. After every action, the NCOs of the respective squads should meet briefly to discuss the winner – if any – and how proficiently the squad performed during the action. Together, the "rival" NCOs will determine the appropriate number of points to be awarded to each squad and mark down the result in their own notes so that the outcome cannot be tampered with. I propose that a score of -5 to +5 be used, where -5 is the score given should a squad be utterly defeated and offer no opposition, and +5 points be awarded for a squad which behaved with great proficiency to overcome their opponents.

For instance, squad 1 ambushes squad 2 by luring them into a glade, deceiving squad 2 that the glade is abandoned and allowing them to become complacent, before ambushing and 'eliminating' the entire unit from all sides. At the same time, squad 3 runs into squad 4 on the road, they briefly skirmish and both retreat in disarray. Squad 1 should receive +5 points, squad 2 lose 5 points, while squad 3 and 4 should receive 0 points.

Control

In both this training chapter and the previous one, there are common problems which could arise: unsupervised Drill Instructors may get carried away, while instructors at MCT could fail to teach what they should or give improper orders. Other than holding solely those individuals responsible, another possible simple solution is administrative. Recruits in Boot Camp and Marines in MCT should be told explicitly, under little stress, and repeatedly, to report what they might consider improper or unfair orders or treatment by their superiors. They should not be penalized for doing so. If a Recruit or Marine claims there's an issue and there is in fact not, but they remain adamant, they should be cycled to a different platoon. They may therefore see that the treatment received by all units is the same.

To implement this protocol, an Officer or NCO should regularly make rounds seeking discrepancies. This individual should have little to no personal relationship to the Instructors. In Boot Camp, each Recruit had an interview with their Series Commander, a Captain, who in essence asked if they were treated correctly (he asked questions such as: do you receive mail, are you abused?). I propose this interview take place with greater frequency to reduce the probability that something bad would actually happen, and to serve as a constant check on the behavior of instructors. I also propose that the interview include a request to report maltreatment of other Recruits in case those Recruits are too uncomfortable or unable to stand up for themselves. I cannot stress how important it is for the chain of command to place more trust its subordinates, and I believe the Corps leans closer to micromanaging its lower levels of command than empowering them as much as it could. I also recommend that further studies on this subject be launched by various commands.

Incentives

One of the major issues I noticed during my time in military training was that there were very few incentives. In fact, the incentives tended to be maligned with reality. Volunteering to do anything was not rewarded because it simply resulted in additional work – not even a 'good job' or 'thank you' after the event is completed. Further, junior Marines in training don't receive any tangible benefit for having been part of an 'honor platoon' in Boot Camp or MCT. The idea is that Marines are expected to volunteer for everything and to accomplish all tasks with zeal. But Marines are still humans, and one of the greatest lessons economics teaches us is that humans respond to incentives.[18]

With this in mind, a small incentive to perform well beyond pride in one's temporary unit should be built into training. I propose that the members of the designated honor squad receive points towards promotion. I reached this opinion by process of elimination. A ribbon or medal would be overly extravagant and inappropriate. An individual award would not be applicable because it is the individual's work as part of the squad that deserves rewarding, which is not the same as the actions of an individual. A letter of appreciation conveys little substance and would be unfitting since the Marine Corps does not issue letters of appreciation for doing one's duty. A few points – say the sum awarded to a Lance Corporal or Private First Class for successfully referring a new applicant to his or her Recruiter – would cost the Corps nothing and provide an enormous incentive for junior Marines to seek out action as part of their squad and thereby become the 'honor squad.' This is especially important since the squad is

---

[18] See for instance Charles Wheelan, *Naked Economics: Undressing the Dismal Science* (New York: W.W. Norton, 2010), Chapter 2.

not actively led by a Noncommissioned Officer, but by junior Marines themselves. Factors that influence initiative taking and creativity must be sought out and implemented.

I also propose that an award be presented to the best NCO. Having spent several weeks in the field away from their usual post in the company of junior Marines is an annoying and difficult task to demand of these NCOs; while the all-encompassing job of teaching and taking care of their Marines is inherently trying. Superior performance should be rewarded with an individual award or certificate of commendation for, say, the top two or three best NCOs. The award cannot be based on the performance of the squad because it could provide a negative incentive to the NCO to actively lead their squad to success rather than letting it lead itself; in other words, there need not be a correlation between the NCO who wins "best NCO" and the "honor squad." On the other hand, it would not be appropriate to ask the squad to vote on the actions of their NCO because it could lead to possibly negative incentives for the NCO to treat their Marines in an unsatisfactory manner to endear themselves. In any case, it would be difficult to parcel out the best NCOs if every squad voted that their NCO was fantastic. The only other possible alternative is that the NCOs themselves vote on who is deserving. Such a decision would therefore be based on both the squad's growth and performance under the tutelage of the NCO, as well as how proficiently the NCO conducted his or her after-action discussions. Furthermore, this policy would incentivize NCOs to conduct their after-action discussions properly and cordially with one another.

Suggestions for Improvement

My recommendations for a change in MCT were made with a focus towards keeping costs as low as possible. Minimal investment would be required to supplement the urban environments that already exist with new facilities, while almost every other aspect of free play combat exercises would be created by the Marines themselves. For instance, a squad may decide to capture the side of a hill, fortify it and use it as a base of operations. Whatever prepared positions, camouflage, or defensive works they had erected would remain there for the next cycles. As time passes and squads shape the grounds to their satisfaction, the map becomes continuously more complicated.

The nature of my proposed MCT is geared towards teaching the fundamentals of being a rifleman on a modern battlefield. But as was shown in previous chapters of this book, Marines may be required to fight in unconventional conflicts. Indeed, as I argued in Chapter 2, I believe the Marine Corps should be pivoted towards those conflicts. MCT as I recommend it would develop baseline competency for conventional operations, the development of judgment and decisiveness for tactics, and provide some leadership experience.

MCT does not necessarily directly develop other traits or provide experience which may be of use in non-combat situations during counterinsurgent operations. For instance, practicing interactions with locals, open mindedness, cultural awareness, going undercover, gathering intelligence, and exercising legal judgment necessary for rationally implementing laws in an occupied area. These elements could be brushed upon by the NCOs of squads as they see fit, but the focus would still remain on small-unit combat. While the MCT I propose does feature significant improvements from its current state, improvements which do overlap with aspects of counterinsurgent warfare, another introductory course could be added – beyond MCT - for every Marine should the Corps pivot to become the nation's preeminent counterinsurgent force. That course could require significant investment to plan and implement.

Part 2

My Boot Camp

Before beginning, some disclaimers should be made about the following material. First and foremost, like the last section, my views and opinions are my own and do not reflect those of the Marine Corps. The Marine Corps in no way endorsed or verified any of the statements made in this book. My training regiment went through the three-phase system. The Marine Corps has since switched to four phases – so training may be somewhat different today, though I doubt it varies greatly. I went to Boot Camp on the West Coast of California, and my experiences may not reflect those of other regiments in California or on the East coast. Above all, the names of individuals, dates, and any numbers herein are fictitious to protect identity and privacy.

I wrote this section in narrative form based on the notes I took in Boot Camp, my memories, and my opinions. In the footnotes of this section are various multimedia links – hosted on YouTube. They render the experiences more lifelike; if a picture is worth a thousand words, surely a video is worth more! I also explain the background of terms, statements, exercises, and elements which are not self-descriptive. The reader will notice that the narrative begins in a very descriptive matter of training and transitions to a story-based narrative. There are two reasons for that. First, I got tired of writing down a chronicle of what we did because it was just so repetitive. No one wants to read (or write) "we woke up, PT'd, went to chow, drilled, and went to bed," for three months of journal entries. Second, it actually mirrors the experience somewhat, for at the beginning of Boot Camp everyone was a stranger that looked the same, and nothing particularly interesting or entertaining happened. As time went by, we got to know one another, instructors loosened up a little, and funny episodes arose. The humor of the environment was dark, sardonic. It may be difficult to understand why something was comical to someone outside the military; what may otherwise have not made any impression had become diverting in the day to day monotony. Looking back on it, I think a big outcome from Boot Camp is the common culture of aggression and humor inculcated into Recruits.

Chapter 1: Phase I

October 11

This is it. My ship date. I'm at MEPS (military entrance processing station) a hospital-like building that serves as the gate keeper for the military. Applicants are processed through medical examinations and paperwork is checked. It's organized chaos, filled with fearful teenagers waiting in lines, rushing between different locations. It would be a perfect place to have a government version of *The Office*. Applicants constantly fail to pass through MEPS because of mistakes in their paperwork or medical discrepancies. Many of the applicants were there for their second, third, even fourth times.

I kept going through a cycle of thoughts in my head. Why did I do this? What was I thinking? But at this point I've burnt my bridges. I don't even have my cell phone. I have nearly nothing left of my normal life. I feel terribly anxious. I must admit I am one of the somewhat fearful future Recruits. The yelling and the stares, the memories of OCS return like screaming ghosts. My stomach tightened at the thoughts running through my mind, the most salient vision was of waking up exhausted to bright lights and screaming in a squad bay.

Having passed that initial processing, we were sent to a hotel near MEPS for the night, we woke up early in the morning to take a bus to the airport. Waiting for our plane, I think about my loved ones. I feel I must push them out of my mind. It's funny how those senses and thoughts of departure affect one. I feel as though there is so much to be said, so much to be shared with them; yet

151

when time was aplenty, I did not pursue such objectives with zeal. In fact, I wouldn't quite know what to say should I see them again. Caught between the weight of the familiar past and a bottomless uncertain future, my mind grabs on to former happier times.

___

October 15
Today is the first time I've been able to write in a few days. We're reaching the 'training days,' whereas before we were in the 'forming days' of Boot Camp. Receiving was terrible. I got off the plane in San Diego. I was thrilled to be going West. Recruits train either at MCRD (Marine Corps Recruit Depot) San Diego or Parris Island. It's a bit of a rivalry in the Marine Corps; for while Southern California is riddled with hills and beaches for Recruits to hike upon, South Carolina is either humid and hot, or covered in rain and snow. Worse still, the area is infested with sand fleas and various other unpleasant critters. I would rather hike up any mountain than endure the crawling feet and bites of innumerable insects under the southern sun.

Like my fellow Recruits, I went to the USO at the airport in San Diego, a sort of lounge area for service members, to wait to be picked up. People had slowly filtered in to the USO from across the nation over the course of the day, the guy sitting next to me had flown in from Hawaii, it had been his first time taking a plane.[19] Though I tried to take a nap, apprehension of what was to come ran rampant in my mind, I was too anxious to fall asleep.

The USO was comfortable, there was free food, plenty of couches, and an incredibly kind staff. In my experience, the staff of all USO's are like that – all

___

[19] Apparently, everything related to the military has an acronym – USO stands for United Service Organization.

wonderful volunteers. As I sat there, I saw a Marine enter bearing the notorious drill instructor hat, strolling in from the dark of the night. The quiet hubbub died away, and everyone sat up straighter, waiting to hear what he would say. Suddenly he barked for us to get on a white bus outside. We boarded and rode with our heads between our knees (so that we wouldn't be able to see the way to or from the depot) as we made the short trip from the airport to MCRD. There were only fourteen of us in the bus that night. We got out and stood on the famous yellow footprints. I heard we were supposed to get yelled at here and experience a miserable introduction to training, but that didn't happen with my group. From what I understand, our platoon was put together last minute using various groups of extra Recruits. My group arrived at 9 PM, so there were fewer games.

Figure 5: The famous yellow footprints of MCRD, with the initial processing building in which we are searched in the back.

We were led into a building, the one with the sign

in the picture above, to be briefed on what we could and could not bring in to the depot, and to make a call home to inform our families that we had safely arrived.[20] I had heard that they would allow religious material in. Always one for pushing boundaries, I brought a huge novel with me the size of a MacBook: Ken Follett's *Pillars of the Earth*, which I thought sounded vaguely religious and could possibly pass the inspection. The various Marines going through our belongings looked at it for a moment and set it back down. I couldn't believe my luck! To see what the contraband room was like, I recommend the link in this footnote.[21]

For the rest of the night we went from station to station to receive gear, sign paperwork, take drug tests, etc. Everywhere we went, we power-walked or drilled. Everywhere we stopped, we stood at attention, and lined up as tight as sardines. It was called standing 'nut to butt.' For the next three days we slept a total of 4 hours because of the amount of equipment we needed to receive and the administrative details we had to cover. I recall they put us in this little conference room and told us not to sleep. Naturally, everybody would pass out. In hindsight, I believe that was the Receiving Drill Instructors trying to work around not being able to give us a real block of time to rest. The guys I was with were all quite convivial and hailed from all over the place. I didn't know just how into the military Texans were, but the large sample size in my platoon quickly remediated my lack of knowledge. All ethnicities were represented, the whole United States coalesced into one platoon.

[20] Though this video features Parris Island, the content was congruent, Thomas2878, "Marine Corps Boot Camp - Contact Your Next of Kin Phone Call," *YouTube*, November 4, 2017, https://www.youtube.com/watch?v=LP2fpn8PRns.
[21] 1200WOAI, "Marine Recruits Arrive at MCRD," *YouTube*, July 26, 2018, https://www.youtube.com/watch?v=YnKPpNArrdg.

I know it doesn't sound that bad in writing, but everyone really hated receiving week. Almost all Marines I've met agree that the worst part of Boot Camp was the two days of receiving – of standing around, no sleep, and the novelty of the depressing new environment. When we went to chow in the morning, we walked down the main road and could hear bangs and screams coming from the training buildings we would soon occupy. Our temporary Drill Instructors were always gleeful when they heard those "sounds of joy," as they called them, greeting the screams with "mmmhmm" and, "hell yeah!"

As expected, we were buzzed bald, a hairstyle aptly named the induction cut. Everyone looked completely the same to everyone else. Getting a haircut was inevitable, and we all knew it would come. But, actually sitting down and having your hair forcefully removed was somewhat of a violent process. It was at that moment that I felt my individuality shorn, to be replaced only by a naked visual commonality. It's strange how much of my personality I felt I lost when someone else razed my hair, definitely an interesting experience to go through. In any case the hair returned and the raw cuts from Recruit training soon become a thing of the past.

On the final day of receiving, my little group packed its gear precisely, for we were moving from our receiving barracks to our training barracks. There was a rumor going around that we would finally be allowed to sleep for some 8 hours that night. Needless to say, we were terribly excited for that order to be given. Out of seemingly nowhere, we picked up a ton of additional Recruits, growing our platoon from 14 to over 80! Worse still, they had not yet packed or inventoried their gear. It took all night to "fix" these individuals. My little group was hence known as the original 14 for the next week or so as we had done a few things the others had not, while the others had received their DOD CAC's– their military ID's - and we had not.[22] By being a little more squared away and having a lesser proportion of bad Recruits than the new group, we felt a little more 'elite' than the new guys. As will be made clear later on, bad Recruits were the people who moved slowly, or were overweight, or yelled softly, or responded poorly, and got noticed by instructors more than the average. It sounds ridiculous, but those little details of being 'more squared away' and whatnot actually made a big difference at that point. We had already begun internalizing the standards of the Marine Corps.

We were introduced to the language of Recruit training during receiving. Most common items had a different name there. I'm not entirely sure what the point was other than to give instructors an excuse to yell at Recruits, because many of these terms aren't used after Recruit training. In any case I thought it was rather amusing.

        Pen = Ink Stick
        Pencil = Lead Stick
        Toothpaste = Fang-Paste
        Toothbrush = Fang-Brush
        Glasses (for sight and windows) = Portholes
        Wall = Bulkhead
        Door = Hatch
        Floor = Deck
        Flip Flops = Shower Shoes
        Sneaker = Go Faster (my favorite)
        Your Head = Grape/Dome-Piece
        Food = Chow
        Cafeteria = Chow Hall
        Hat = Cover

---

[22] DOD CAC stands for department of defense common access card.

Bed = Rack
Room in which we slept = House/Squad Bay
Bathroom = Head
Use the bathroom = Make a Head Call
Beyond translations, a few terms are also in order:

Prac: Prac was 'knowledge.' Recruits were given a set of ditties, terms, and statements to learn – it was called knowledge. Examples of knowledge included the chain of command or how to wear your uniform correctly. Prac was a subset of that set of knowledge which we learned by screaming it out again, and again, and again. The instructor would yell a prompt, and we would scream back the correct answer or ditty. For instance:

Instructor: "First Female Marine!"
Recruits: "First Female Marine Aye Aye Sir, Opha Mae Johnson, 1918!"

Instructor: "Battles of World War One!"
Recruits: "Battles of World War One Aye Aye Sir, keyword BBS, Belleau Wood, Blanc Mont, Soissons!"

The most complicated one was the medical ditty for saving a fellow wounded Marine,

Instructor: "54321!"
Recruits: "5, five types of wounds, keyword APAIL: avulsion, puncture, abrasion, incision, laceration; 4, four life-saving steps: start the breathing, stop the bleeding, protect the wound, treat for shock; 3, three types of bleeding keyword VAC: venous, arterial, capillary; 2, two types of wounds: open, closed; 1 [this part of the ditty was for the T-55 day test when we have to perform a first-aid scenario on a dummy], pick up the

157

weapon, up down, left right, all around clear, combat mindset, low crawl to the victim, are you alright, are you okay? You call the corpsman, triage the victim, head to toe, toe to head, don't forget the arms, expose the chest, treat for shock, pick up the weapon, sight in, mission accomplished Aye Aye Sir!"

That entire ditty was memorized and screamed for a seemingly infinite number of times, everywhere we went. We would get in screaming contests with other platoons doing that ditty. As a result, years away from Boot Camp, I still remember every word of it. Prac was perhaps the worst part of Boot Camp other than receiving week. We sometimes screamed 'knowledge' for hours on end, the exact same thing again and again. It ensured that these concepts would be ingrained within us, so as much as I hated prac, I understood its purpose.

Prac ap: short for practical application, in Boot Camp this term was typically applied to the portion of the T-55 Day test where we had to apply medical aid to a wounded dummy.

Prac Hat: This is the Drill Instructor responsible for teaching us knowledge and being prepared for the T-55 Day test. Though the responsibilities of instructors overlapped heavily, they also had a specialty. The prac hat would ensure we took practice tests and received the 'knowledge' we needed as basic Marines.

Drill/J Hat: The lead instructor for teaching us to drill, on the East coast they're known as the 'Heavy.'

Kill Hat: This instructor is responsible for discipline, and everything necessary to enforce and develop that discipline. When one typically thinks of a Drill Instructor, the kill hat comes to mind. He is the one constantly making corrections, enforcing discipline, leading IT sessions (explained later), and generally on the lookout for any possible infraction. In my platoon, at the end of first phase we received two new drill instructors who partially took over the role of kill hats while our previous kill hats transitioned to become the prac hat.

Senior Drill Instructor: As his name implies, the Senior Drill Instructor is in charge of the other Drill Instructors. He is essentially responsible for our overall treatment, that instructors follow the guidelines of the Marine Corps, and that we receive the training we ought to. The Senior Drill Instructor earned his place, having trained multiple cycles of Recruits. Many Enlisted came from broken backgrounds, and some came to see the Senior Drill Instructor as a father figure. He was physically with us far less than the other Instructors because of the "back-office" work he had to do. He distributed mail, sometimes let us eat sweets at chow, answered our questions, and was always available to discuss personal problems.

PT: Physical training, it consisted of running and calisthenics.

On Line: Recruits are either PT'ing, eating, running, drilling, or standing still. If they stood still, they stood at attention with eyes fixed ahead. When inside the squad bay, we stood with our heels touching our footlockers in a perfectly straight line, at the position of attention (POA). Your body was fixed, head and eyes staring straight ahead, and fists clenched at your sides, thumbs on your trouser seams. That collective position was being 'on line.' Being on line allowed instructors to see that everyone was essentially on the same page, that a task had been accomplished, that everyone had the right equipment, etc.

T-55 Day Test: This was a test which took almost an entire day to do. It served to ensure that we had learned all the basic knowledge of a Marine. T-55 signifies training day 55, the day the test would take place within the training cycle. All the prac and prac ap we did was aimed at passing this test.

Games: How we spent most extra time in Boot Camp, called fuck fuck games but not in this book for decency's sake. Games were a form of training, really a group punishment which always involved screaming along with variations of standing up and sitting down, or lifting an item and holding it out, or picking something up and putting it down quickly. They could get more elaborate and of greater severity, the worst ones involved dumping our possessions and mixing them around, throwing everything in the showers, redoing our carefully made racks, or just straight up moving our racks around and messing up the meticulous order of the squad bay – which of course we had to fix later.

After everyone's gear was finally squared away, we were given a few hours to sleep, but not the anticipated 8 hours. We woke early and were all marched over to our training squad bays,[23] laden with our gear in the newly issued sea bags (duffel bags) and war bags (a smaller duffel bag). We were assigned to Platoon 1025.[24] We had four days of 'forming,' which involved being taught military customs, courtesies, and an introduction to drill. There was a procedure for everything. Military customs at Boot Camp consisted of yelling a proper greeting when a Marine was near, which could be either "Good morning Sir, Good afternoon Sir, or Good

---

[23] When we met our Drill Instructors, they gave an introductory speech. A good representation is available at the following link: AiirSource Military, "Drill Instructor Gives EPIC Speech – United States Marine Corps Recruit Training," *YouTube*, June 2, 2016 https://www.youtube.com/watch?v=-Ns2FkZNTC0.

[24] Again, as I wrote earlier, all identifying information including the number of the platoon and any names are fictitious.

evening Sir."[25] Since we had no watches, the greeting was based on the last chow eaten. If one had eaten afternoon chow (lunch), it was afternoon – even though it might be 10AM, you would have had no idea. The calendar and days likewise disappeared from knowledge. I quickly lost any idea of what day it was – often even what month we were in. The proper greeting, like everything else said in Recruit training, was screamed. When the platoon passed a Drill Instructor, there was a cacophony of "Good morning Sir!"

We were taught to clean the barracks, how to sweep, swab (using towels attached to a broom to 'mop' the floor quickly), shave, dress, make a bed, etc. Every conceivable aspect of basic living received a period of instruction.

Above all, Recruits had to respond to everything said to them – almost always with a "Yes Sir," "No Sir," or "Aye Aye Sir" (the dual Aye coming from the naval background of the Marine Corps). The responses were screamed with enormous effort, and platoons competed for volume. The idea was that one could tell a good platoon from a bad one by how loud it was, how well it drilled, and with how much discipline it moved or stood still. I can't stress that Recruits always had to respond; in one way or another they acknowledged the receipt of an order or responded to questions.

The following were a few of the more elaborate procedures that were learned over the course of receiving and the early part of first phase. These 'procedures' were performed, for three months in a row, every single day.

---

[25] At that time, there were no female Recruits or female Drill Instructors at MCRD San Diego. Women all went to MCRD Parris Island for their basic training. Had there been female Drill Instructors, we would have said Ma'am instead of Sir.

I: A Recruit is not an individual, he or she is a Recruit. A Recruit cannot refer to themselves as I or me, and items cannot belong to the individual, they belong to the Recruit. Therefore, it is not I, it is this Recruit. It is not me, it is this Recruit. It is not my backpack, it is this Recruit's backpack. The Drill Instructors got very excited when Recruits forgot about their status!

Chow: The Company or Series (a Series was half a company, my platoon was in 'follow Series,' the other half of the company was called 'lead Series') was given a specific time to be at the chow hall. If Drill Instructors thought their Recruits had been behaving well or had a particularly busy day and needed to save as much time as possible, they would race to get to chow. The earlier one arrived at chow, the more time the platoon had to eat. Of course, that was only to a certain extent, for other platoons and other companies had to pass through as well and competed to arrive early.

We assembled in front of the squad bay in formation, then drilled over to the hatch (door) that was assigned to our platoon. We briefly stopped at some outdoor sinks – wash racks – to wash our hands and continued to our destination. Once we arrived, if there was a long line, we either continued to drill or 'formed a school circle.' If there was no line, we became the line. A school circle was a tightly packed group assembled in front of a Drill Instructor so they could teach something. Typically, at first, the Instructors discussed mostly drill, told us we were all bad Recruits, or demanded prac. We would all be utterly quiet, waiting for a prompt to scream. Otherwise we did our best to not be noticed. Later on, in third Phase, school circles became more relaxed as they allowed us to ask questions about the Marine Corps.

We lined up – nut to butt, naturally – and prepared to enter the chow hall. Thereupon the Guide posted himself in front of the platoon and screamed:

"Ears 25!"

The platoon answered, "Open Guide!"

"Ears 25!"

"Open Guide!"

"The platoon will be sitting on Drill Instructor highway, does the platoon understand?"[26]

"Yes Guide!"

"Does the platoon understand?"

"Yes Guide!"

Thereupon the Guide yelled: "Attack!"

"Kill 23!" screamed the platoon, referring to the other platoons in our Series (which was, remember, our half of the company).

"Attack!"

"Kill 24!"

"Aaaaattaaaaack!"

"Kill Kill Kill them all!" everyone screamed.

The Guide would thereupon post up at the hatch and call Recruits in so as to regulate the length of the line, while the platoon continued to do prac or was kept quiet and disciplined by the Recruit Squad Leaders. The moment a platoon of Recruits was left unsupervised by Drill Instructors, the Recruits began talking and making an extraordinary ruckus. More problematic, this lack of discipline would be seen by other Drill Instructors, which would reflect poorly upon our Drill Instructors and Series in general – meaning no lapse in discipline could be tolerated!

---

[26] The location in the chow hall could be the drill instructor highway, the far bulkhead, near bulkhead (remember bulkhead means wall), and the salad bar.

When we entered the chow hall and received our food from the kind chow hall workers, we power walked to our designated area, set our trays down, and grabbed a drink, either Power Ade or water. In the chow hall, one had to power walk at all times and (naturally) scream the proper greeting to any Drill Instructor in the vicinity. One had to place one's canteen (a Recruit had some sort of water source on them at all times) under their chair, touching the pole. You yourself had to be sitting up straight, left hand on the left knee, your chin touched your chest as your right hand furiously forked food into your mouth. The cup touched the tray and was placed directly in the middle of the tray.

You were given plenty of time, maybe 5-8 minutes, to eat if you were one of the first to sit down. Coming in as one of the last Recruits throughout Boot Camp, I was always rushed to eat – less than 2-5 minutes. Nevertheless, I rarely felt hungry, and do believe we were well treated by our Instructors – though it certainly didn't feel that way at the time, especially on the days we were pressed for time and left almost the moment we sat down. In what would later become one of the platoon's favorite jokes, one of our Drill Instructors – Sergeant O'Brien – often said with disappointment "we're still here 25... we're still here..." just as the last Recruits were about to take their seats.

As we ate, the Instructors continued yelling at us. Because they didn't want us to choke on our food as we ate, they had us stomp on the deck instead. In what was, in my opinion, hilarious, Instructors often ended their comments with "stomp, stomp" and Recruits furiously pounded the floor over and over with their feet to acknowledge whatever statement was made to them.

In the eventuality that you finished your drink and had time to get another drink, or at the end of our time for chow, you stood up by your seat and screamed the proper greeting for the chow hall, "Good morning [or afternoon/evening] ladies, good morning [or afternoon/evening] gentlemen!" Most of the time, especially if the Instructors didn't think you were a perfect Recruit, they screamed "Bullshit! Get back down." Or, "No volume, do it again," or "Yeah right!" or some variation of the order to sit down and scream again.

When chow ended, we ran out, formed up outside, and drilled back to our barracks. There was never extra time to eat, but there was *always* time to do pull-ups. Outside almost every Marine barrack and chow hall, there are pull-up bars. We crushed out pull-ups for a few minutes before running back into the squad bay for a head call (to pee) and continued our day. Almost all of us became quite good at pull-ups. I came into Boot Camp doing 21, and left being able to do 32.

Cookies: The subject of sweets at Boot Camp was so enormously important it requires its own subsection to address. Months out of training, some Marines still spoke about it. You see officially, Drill Instructors were not allowed to restrict us from eating what was available at the chow hall. We were technically allowed to grab any of the sweets that were available on the table. On the other hand, at least one Instructor from every platoon currently in the chow hall stood at the sweets section, which also happened to be the end of the line to receive food. That allowed them to see that Recruits were getting the proper food and nutrients every day - enough vegetables, fruit, carbs, and protein. It also allowed them to note if they had the audacity to stop and grab a cookie. Of course, when passing the group of Instructors, one had to scream the proper greeting before continuing on, which was a great time for them to tell you what they thought of you at that moment. For example, in the evening, I would almost always get told I was a shitbag Recruit because my facial hair had grown into a five o'clock shadow (I had thus lost my clean shave).

Anyways, the first couple Recruits that grabbed a sweet quickly learned it was a bad idea when they got IT'd right after chow to work off those extra calories. Eventually no one grabbed any cookies. That wasn't any fun at all. So, the Instructors would mix it up a bit. Some Instructors would order Recruits to get sweets and watch over them to see if they would eat it or not (most did not, though I did see a Recruit hide a cookie at the last minute instead of throwing it away). Naturally if you did eat it, you would receive some additional training. Some Instructors really cared if you got sweets, some did not. One of our Instructors in particular really cared, Staff Sergeant Esposito. He cared so much we started calling him the Cookie Monster. His obsession with cookies will indubitably come back up in this book.

The way people got sweets is when they went to the chow hall on their own, typically because they were going to dental, medical, or as prayer leaders on Sunday before religious services. There was however always the risk that you would be caught. I never ate a sweet without constantly looking around like a scared deer, but I promise you those were the best cookies of my life.[27]

BDR: BDR stood for Basic Daily Routine. Every night we did this routine to the t. We got on line in our white cotton briefs, which were called panties (more colloquially known as tighty-whities) and shower shoes. In the 'panties' would be tucked a T-shirt and skivvy shorts, along with one's 'money/valuable bag'. This bag contained, unsurprisingly, your money and whatever

[27] No offense intended to you mom!

small valuables you kept with you in Boot Camp, like a Social Security Card. I kept my wallet and keys in there.

We would all be lined up and the Guide would pound with all his strength on the Drill Instructor hatch (the door to the room where they slept and hung out), he then screamed, "Plaaaatoooooooon!"

Thereupon the platoon yelled, "1025, locked, cocked, and ready for hygiene inspection!" The Instructors often requested the Guide repeat this a few times because he or the platoon weren't loud enough or because some Recruit was moving or looking around.

Finally, the Instructors walked out and yelled 'VIP's!' The platoon would then yell "VIP's!" The Squad Leaders, Guide, Scribe, and those who needed to go to medical would step forward and yell, "Aye Aye Sir, Aye Aye Recruits, carry on Recruits!" To which the platoon responded, "Kill!" The group of 'VIPs' assembled in a line and were inspected by the Drill Instructors. Then one Instructor said, with a dramatic sweep of his arm, "Hit it!"

We raised our hands and forearms parallel to the deck while keeping our elbows pinned and turned our heads to the right (so they could inspect the hands and sides of the head), while saying, "Snap! Good evening Sir, Recruit [last name], 0311 infantry regular!"[28] The Instructor walked up and down and noted physical infractions or wounds. I was personally never badly injured but sometimes had difficulty scrubbing the stain of ink out from under my fingernails; I almost always had a five o'clock shadow by this time and was yelled at for those failures. Sometimes failing this inspection

---

[28] The 0311 signifies your MOS, your military occupational specialty. Though none of us yet had an MOS because we had not finished basic training, we would all be considered basic riflemen after graduating Boot Camp. Saying 0311 was perhaps a way for us to become comfortable with that.

would mean one would be assigned fire watch (sentry) duty – though sneakily, our Scribe often didn't actually write names down.

His first inspection done; the Instructor yelled "flip!"

"Pop, this Recruit has no personal or medical problems to report at this time."

"POA!"

"Turn."

"To the right!" Whereupon we turned around in a clockwise manner.

"Right foot."

"Right foot Aye Aye Sir!" We raised our right foot so that he could inspect underneath it.

"Left foot."

"Left foot Aye Aye Sir!"

"Post!"

"Post Aye Aye Sir, good evening Sir!"

"Get dressed."

"Aye Aye Sir!" We ran back to our places and got dressed, held out our money/valuable bag in front of us as the Instructor repeated the same process down the line to inspect the entire platoon in small sections.

When finished, the Instructor yelled, "Count!"

We responded, "Grab!" and grabbed our money/valuable bags.

"Off!"

"Snap!" we yelled as we held it straight out in front of us. The first Recruit started the sequence and yelled "one," then lowered his money/valuable bag, and so on until the Guide who screamed his number – for instance eighty.

"Aaaaand eighty!"

The platoon yelled, "Good evening Sir, the count on deck is eighty United States Marine Corps

Recruits!" If the count was not perfect, we would do it again and again.

Bedtime: After BDR, in later phases, we were given one or two minutes to pray. After prayers, we stood back on line, and – in Phase III, an Instructor would say, "Turn on the radio!"[29]

We responded with, "click click mmhhhmmmmm" then chanted the Marine Corps hymn. I remember I once asked if we should try to harmonize while singing or to yell as loud as possible. The Instructor responded, "Both!"

After singing, we held out a full canteen in front of us.

"Prepare to hydrate!" exclaimed the Instructor.

"Prepare to hydrate Aye Aye Sir!"

"Do it now, move."

"Kill! 1 pop stop 2 pop stop 3 pop stop 4 pop stop 5 pop stop 6 pop stop 7 freeze!"[30]

"Drink 'em!"

"To the Marine Corps!"

We chugged the entire canteen, which was actually annoying because it was a ton of water, and the moment you would otherwise have gone to sleep you had to get up to use the head.

"Cap 'em!" a moment went by as we screwed on the lid of the canteen. "Secure the canteen!"

"Secure the canteen Aye Aye Sir!"

"Do it now, move!"

"Kill!" We slammed our canteens against our footlockers and placed them delicately on its surface.

---

[29] In Phases I and II we omitted this step for we didn't yet rate singing the hymn.
[30] This ditty is in reference to the movements made with the rifle during the drill order inspection arms.

After hydrating, we received the command to move our footlockers one yard away from our racks to make an opening so that fire-watch could clean that area. We also set up our boots and go fasters on the footlocker in a neat manner, possibly so Instructors could ensure no one had lost their equipment.

"In between the racks, do it now, move!"

"Kill!"

"Prepare to mount!"

"Prepare to mount, Aye Aye Sir!" We raised our arms above us, then slammed our arms down on the top rack with all our strength while screaming, "Discipline through pain!"

"Mount!"

"Kill!"

"Crunches until lights." We did crunches in our racks until the lights were turned off, usually for exactly 2 minutes.

Once the lights went off, we got into a position of attention in our racks and waited for TAPS to play (a melody which is played in honor of the dead) after which we waited a few more minutes before being allowed to use the head and enjoy our night.

Nighttime: The night was more or less free time, though Instructors still ran around causing mischief and making sure Recruits weren't up to no good. It's at night that you did your laundry after first Phase, wrote letters, prayed, whatever it might be. I read *Pillars of the Earth* under my sheets. It was also coincidentally a time when weird and hilarious events tended to occur, as will be seen later in the book.

Fire-watch: At every hour of the night, there were at least four Recruits acting as fire-watch; a fire-watch is basically a sentry. One of them was placed in front of the main hatch to report on the status of the platoon and take note of what was happening, while another was at the back exit to make sure no one went in or out the back entrance. Fortunately, while nothing bad happened during my time in Boot Camp, fire watch is absolutely necessary to ensure that unacceptable events didn't occur or to mitigate those events as best they can. For instance, fire watch would deal with a wild animal attempting to enter the barracks or a Recruit going suicidal. The other two Recruits on fire watch were 'rovers' and typically cleaned (though many snuck away to write letters or to sleep in secret). Cleaning consisted of mopping the head and the squad bay, reorganizing the cleaning supplies, and wiping down surfaces.

Fire-watch was responsible for turning the lights on and off every day, which they did by announcing beforehand, "Five minutes to lights, five minutes to lights, five minutes to lights!" for every minute, then every 30 seconds, and so on. Then, "Lights, lights, lights!" as they turned them off or on. The timekeeper was a Drill Instructor in a Drill Instructor hut somewhere, and the time to lights was repeated by Recruits around the building, so we heard a cacophony of Recruits screaming random time intervals. We often heard them scream 4 minutes till lights, then 8 minutes, then 30 seconds as the telephone game was played across the complex of buildings.

Fire-watch was also a good way for Instructors to continue messing with us at night. They would typically walk in and scream "fire-watch!" over and over until some Recruit ran over, and they would tell that Recruit to do something ridiculous like pick up a piece of lint off the ground.

Most relevant of all however was front post. Every time an Instructor walked in, the front post Recruit had to snap to attention and 'report his post.' The full report went something like this:

"Good evening Sir (or gentlemen if multiple Instructors walked in), Recruit [name] reports building 1022 squad bay 3 all secure. The count on deck is 80 United States Marine Corps Recruits, 80 M16A4 Service Rifles, all double locked and secured in the armory, and 80 United States Marine Corps footlockers, all properly secured. There is nothing unusual to report at this time, good evening Sir!"

There were tons of ways a Recruit could mess this script up; not saluting, not saluting well, not saying the proper greeting, reporting without actually having made a count of equipment being reported, messing up the order of items to report, messing up the report generally. Instructors messed with the front post guy all the time, making it a very unpopular station to hold.

Common Orders and Responses:
Eyeballs = Click Sir! (you look directly at the Instructor)

Zero = Freeze! (freeze your body, in mid-air if you have to!)

Ears = Open Sir! (listen, pay attention)

Open your [freaking /big/fat/sweet ass] mouth/face/lips = Aye Aye Sir! (get louder)

No Volume, Sprint = Aye Aye Sir! (run because you weren't being loud)

5, 4, 3, 2, 1 you are = Done Sir!

Do it now, move = Kill! (this was placed at the end of an order, upon saying kill the order would be executed)

War cry! = Ahhhhh! (just screaming)

There were three phases to training, which are somewhat self-explanatory. Phase I revolved around basic Marine concepts and predominantly featured games, drill, and classes. Phase II involved learning to fire a rifle and developing field skills such as navigating with a compass. Phase III was the consummation of the process where lessons and skills were refined, and we were tested on what we did in Boot Camp. After the crucible, we became Marines and had a week (Marine Week) dedicated to administrative matters plus a few more classes. A good visual synopsis of the entire process can be found in the footnote.[31]

---

[31] Foxtrot Alpha, "Training Summary: 13 Weeks of Marine Boot Camp - Recruit Training at MCRD San Diego," July 12, 2017, https://www.youtube.com/watch?v=QTZDwTCYis0&feature=youtu.be.

Phase	Week	Monday	Tuesday	Wednesday	Thursday	Friday	Saturday	Sunday
Arrival Week	Pick Up	Forming Period				Pick Up/IST	In-House Procedures	Religious Services
Phase I	1	Intro to M16	Interior Guard	First Aid	MCMAP	Marine Corps History	PT	Religious Services
	2	MCMAP	Pugil Sticks I	PT	Obstacle Course	MCMAP	PT	Religious Services
	3	Confidence Course	PT	Pugil Sticks II	PT	SDI Inspection	Initial Drill	Religious Services
	4	Swim Week			Pugil Sticks III	Marine Corps History	Movement to San Diego	Religious Services
Phase II	5	Grass Week					Hike	Religious Services
	6	Table I					Hike	Religious Services
	7	Field Week			Table II		Intermediate CFT	Movement to MCRD
Phase III	8	Team Week					Intermediate PFT	Religious Services
	9	PT Week					Sustenance Hike	Religious Services
	10	Final PFT/ Clothing	MCMAP Test	Final PFT	PRAC Test	Rappelling	Company CO Inspection	Religious Services
	11	Gas Chamber	Crucible		Reaper Hike	Gear de-Issue	Movement to MCRD	Religious Services
Marine Week	12	Orders Brief	Battalion CO Inspection	Graduation Practice	Motivational Run/ Family Day	Graduation		

Figure 6: The three-phase training system matrix for Boot Camp, reproduced from the publicly available matrixes available online; in theory each block identified the most important activity of that day, in reality many activities were mixed around and the majority of days were more or less the same.

174

October 18[32]

After morning chow, the platoon went to MCMAP. As with every activity in Boot Camp so far, we spent more time yelling and being punished than doing the exercises and learning. MCMAP is the Marine Corps Martial Arts Program. It taught us the fundamentals of hand-to-hand combat. As one rises in rank in the program after Boot Camp, or 'belts up,' MCMAP apparently becomes more interesting. The first belt is the tan belt, which is earned upon successful completion of Recruit training. My Company earned its belt less by hand-to-hand combat, and more by doing 'inchworms,' a dynamic exercise that strengthens the core and shoulders. We did inchworms and war cries the majority of the time in a big pit made of tire pieces, which Instructors called the dojo. A link to a video synopsis of tan belt MCMAP is in the footnote.[33]

We drilled for some time after that, then got afternoon chow. Fourteen of us had not previously had our photos taken for our military IDs, so while the rest of the platoon sat in the 2nd or 3rd class of the day, we went to get IDs.

I was bitterly disappointed because the platoon I was in seemed terribly shoddy. A large portion of the platoon was not motivated to be there. A sizable number of Recruits were clearly frustrated with the situation, as were our Drill Instructors. I worried that Phase I would drag on and on, and that I would learn nothing at all as the platoon failed to gain any discipline until the very end

---

[32] Dates are shaky in Phase I because, as mentioned earlier, I had absolutely no way to know what day it was; these are just approximate dates.
[33] US Military Videos, "Alpha Company MCMAP," *YouTube*, November 5, 2016, https://www.youtube.com/watch?v=OZ9qUPv-KXQ.

of Boot Camp; I feared we would play games for the next three months.

MCRD was an interesting place. We walked among buildings three stories high, all the same architectural style, all the same dull yellow. Most buildings were a series of squad bays connected to one another, and each squad bay contained a single platoon. They were more or less large blocks of cement pierced by windows, from which echoed screams. The stairs were exposed to the outside but shuttered behind bars; bars had been installed to keep Recruits from trying to commit suicide by jumping. Silhouettes of Drill Instructors could be seen lurking in ever shadow. There were no trees nor grass. The entire area was covered in dust, dirt, and rock with only a few courageous bushes growing beyond the buildings.

Other than the squad bays, we regularly visited a few other locations on base. Most prominent was the chow hall. In front of it was a large drilling space. The outside was layered with arches under which Recruits waited for their turn to be processed through chow. Inside, it was just a cafeteria except the seats were attached to the tables, and the backs of the chairs were labeled with motivational labels for Marines like Leatherneck and Devil Dog. The military loves uniformity, and the base followed those principles in its physical appearance – all buildings were the same material, color, and architectural style.

A few minutes away from our squad bay was the parade deck. It was an immense space, a huge flat expanse of blacktop upon which we underwent our drill tests and where the graduation ceremony took place at the end of Boot Camp. The perimeter was ringed by various bureaus and office spaces, lovely low buildings that looked more in keeping with a repurposed Spanish mission, like you would see at Stanford, rather than the sad gravelly world only paces away. Alongside the parade deck was an imposing amphitheater. It is often seen in photos of MCRD San Diego. We had a few briefs and classes in there as a Company. However, almost all our formal classes were held in a different building off the main road, a few minutes away from our squad bay. The classroom itself would have fit in quite well in a college or a large high school.

Finally, the last building we had access to was the PX, which was a small store that sold everything that a Recruit might find necessary during Boot Camp, from socks to razor blades. The PX being next to the barber, we ended up making a trip about once a week to get a haircut and to restock our personal supplies as well as anything the platoon needed (usually cleaning supplies). That trip was known as making a "PX call."

———

October 19

It was interesting to see the changes happen: those who started highly motivated were still motivated and were becoming well assimilated, while those who seemed to possess average motivation were pushing through.[34] But the poorly motivated were quickly exposed as such and brought down the rest of the platoon to their operating tempo. As a result, they held us back.

At all times we were led by our Drill Instructors. In total we had two sets. During receiving we had two Drill Instructors who passed us quickly to our new, permanent ones. They all wore either cammies or more formal stiff collar tan uniforms. But regardless of their uniforms, they always wore a large brimmed hat called a campaign cover. From what I've gathered, this cover

---

[34] "Motivation" refers to how quickly an individual moves, whether they volunteer for tasks, if they scream loudly, and if their general disposition is ready and alert rather than beat down and tired.

originally came from the professional instructors on the shooting ranges to shield them from the elements during their long days outside. That hat immediately conveyed a sense of fear and apprehension upon the gazer. The Drill Instructor was a walking fury.

As hard as Boot Camp could be on Recruits, it was no doubt harder on the Drill Instructor, for they were nonetheless human, they worked longer days than we did, and exerted themselves past the point where most would have collapsed from exhaustion. When we hiked, they ran up and down the line screaming continually. When we were IT'd, they jumped up and down, waved their arms frantically, screamed orders, and feverishly sprinted all about. Moreover, while we only stayed in Boot Camp for three months, they were assigned to their post for years. I can imagine that fulfillment must not be easy for them either: the moment their Recruits began to exhibit discipline and to work as a platoon was when they graduated, and the Drill Instructor started all over with a new batch of 'lost, disgusting civilians.' A truly Sisyphean task.

Their full titles were somewhat strung out. It consisted first of their billet or job – either Drill Instructor or Senior Drill Instructor – then their rank and last name. They were all clean shaven and nearly completely bald at all times. They didn't have to remove their hair, they chose to, in order to look the same as one another and fit in better with Recruits. One of the Leadership Principles of the Marine Corps is to lead by example, and they always did so. Our platoon started with three Instructors. The Senior Drill Instructor was Senior Drill Instructor Staff Sergeant Sheehan. He radiated toughness, a perpetual frown graced his face to such an extent that were a smile to arise, it would appear unnatural. He was a decorated veteran, having served in many campaigns and experienced hardships most of us would never dream of. From the very moment we saw him, we had nothing but respect and esteem for this man.

Staff Sergeant Sheehan was assisted by Drill Instructors Staff Sergeant Esposito and Sergeant O'Brien Esposito was tall and thin, with the most incredible thick eyebrows that curved upwards and lent his face both an eternal surprised look and a certain kindness. A smile was always lurking under his cover, waiting to break out at any opportunity; Esposito was mischievous. His voice gave out early on in the cycle, replaced by a husky bark that was often hard to discern and understand, as a result he never really yelled. In fact, it was a rather important issue because when the platoon was surrounded in noise, we would lose his voice. Our order would disintegrate, and chaos ensued, in particular during Drill.

He was our Drill Hat and very much saw himself as a teacher, dedicated to demonstrating good leadership and teaching us the requirements of being basic Marines. In contrast Sergeant O'Brien started as both our Kill Hat and Prac Hat. A few weeks into Boot Camp we were joined by two new Drill Instructors: Sergeant Reyes and Sergeant Duarte, who had only just graduated from Drill Instructor School and both became Kill Hats, though Sergeant O'Brien never really seemed to 'lose his roots' from that billet and remained a pit-bull to the very end. Sergeant O'Brien was shorter than the previous two Instructors, clearly of Irish stock. He often became incredibly angry. His face became engorged as it turned red and was crisscrossed by thick veins of fury. One of his favorite ways to show how pissed off he had become was to remove his cover, look down at it with disappointment, then give his command with a renewed note of urgent fury. We all thought he had a tendency to get a little carried away, and that he had to be checked by the other Instructors. I recall one instance when our

Senior Drill Instructor came out to calm down the games after he heard Sergeant O'Brien give the order, "Racks outside, platoon formation, do it now, move." He was greeted by the sights of a few Recruits desperately trying to jam their racks into the windows like babies attempting to force a square block through a circle opening in their shape-sorter. Most of us just hoisted our racks in confusion.

We almost exclusively wore our newly issued cammies and 'go-fasters'. Our cammies were fully buttoned up to the neck, and our go fasters looked ridiculous. I thought it was because they wanted us to be identified as Phase I Recruits, and to earn the right to wear the uniform correctly, as we would in Phase II and III. It was also because we spent so much time on our feet and having boots would have increased the likelihood of getting shin splints and other injuries. The sneakers allowed us to ease into shape a bit before boots were required. In any case, we looked terrible. We spent the majority of the day playing games, getting dressed and undressed over and over to a count, putting things in bags and taking them out, lifting bags and canteens up and down, up and down, up and down, repeatedly. Most days passed without anything happening of note, we played, had chow, returned to the house, and played some more.

If a Recruit made a mistake or caught an Instructor's attention for almost any reason, the Instructor would blast them, or chew them out - which consisted of screaming as close to their face as possible. In the past, Instructors had been allowed to hit Recruits, but that was no longer tolerated. Sometimes, multiple Instructors would converge upon a Recruit and all blast him together.[35]

---

[35] For a more holistic view of this concept, I recommend the reader watch the following video on YouTube: Compiler Revolt, "ULTIMATE US

As I walked around during the day, I felt alright but the moment we took a break – say for a lesson – I wanted to pass out immediately. Sleep almost literally blanketed my eyelids. One of the purposes of Boot Camp was to constantly stress individuals and keep them moving. Through constant exposure people become better at handling loud noises and tough environments. Your mind dulls or blocks out some of the noise, the din of shouting becomes normal, and your body adapts to the seventeen-hour days of continuous movement.

––––––

October 20

We woke up before the sun rose, quickly dressed and lined up outside to march to chow. We drilled a bit after chow, then went to a class on first aid procedures during combat. Next, we PT'd. For PT we ran outside in green on green (green shirts and green shorts) along with our canteens. We ran to a spot a few meters away from the house (the squad bay) and formed up as a Series to wait as the Instructors went over safety briefs and practiced a few emergency protocols. Next the entire Series ran a few hundred meters to the area in which we would PT to carry out the plan. In this case it was a set of runs, then we cycled through the various workout stations such as pull-ups, dips, and curls. We formed up again, ran back to the squad bay, and took a "shower," which was actually a running line through the showers that barely got one wet. We changed into our cammies and went to chow for lunch.

After lunch we drilled more and went to another lesson on first aid. Exhaustion was right around the corner. My threadbare stamina, sustained only by constant movement, was a hallmark of Phase I.

At all times of the day, we were observed by the Drill Instructors. They noticed every detail: a hair that escaped from the razor blade, an unfocused gaze into the distance, a button missing from a blouse, a misplaced thumb while standing at attention. If we weren't running, sitting, or drilling, we were standing at attention: feet together at a 45-degree angle, shoulders back, hand in a ball with thumbs on the trouser seams, head and eyes straight to the front. While we were on line or in a school circle listening to what Staff Sergeant Esposito had to teach, Sergeant O'Brien continually inspected us to ensure we stood correctly.

We were introduced to IT today. IT was an intense version of a game where one screamed and switched between exercises consisting of high knees, sit-ups, pushups, jumping jacks, mountain climbers, and steam engines while war crying at the top of one's lungs. Those unfamiliar with the process can find it on YouTube. Apparently, it is highly regulated by the Corps for fear of excessive hazing; it was used for correcting minor disciplinary infractions. The order to get IT'd was usually one of the following: 'go to the quarterdeck,' 'go die,' or 'empty your pockets.'[36]

October 21
After lights we went to chow and had a MCMAP class. After MCMAP, we did the official Marine Corps obstacle course.[37] The course was identical to the one at

---

[36] See for instance: DevilDog, "Getting Smoked by a USMC Drill Instructor." YouTube, March 14, 2015, https://www.youtube.com/watch?v=kwQPJu228Kg.
[37] Visit this link or type "USMC Obstacle Course" on YouTube to see what the course is like, Greg R, "USMC Obstacle Course," *YouTube*, May 5, 2013, https://www.youtube.com/watch?v=NNc7cvpd5qk.

OCS. In fact, the course is the same across the entire Marine Corps. However, there was little pressure to perform here so it took a very long time to get through; at OCS we sped through the course multiple times in just a few minutes. Here the speed was set by the slowest Recruit while the multitude lined up to wait their turn. The course could be rigorous, involving rolling over bars, jumping over logs and walls, and climbing up ropes. Under pressure, it was exhausting; here with no pressure to complete the course quickly, it barely took any effort.

During MCMAP we learned a few elbowing and punching techniques. Afterwards we drilled for a while, got chow, drilled again, and went to another class. The Instructors demanded yelling and little else. The only Instructor who didn't act that way was Staff Sergeant Sheehan. He seemed to want us to think on our own. Training was really all about showing discipline and following simple commands. That was it so far. Hopefully it would be better later on.

In class, the teacher spent more time ordering us to be loud, to not fall asleep, and to address them correctly, which meant standing at attention and screaming: "good morning/afternoon Sir, Recruit [Name], Platoon [platoon number]!" All the other Recruits would exclaim an unmotivated "Kill!" after which you screamed whatever you needed to respond or say during the class.

The only time we saw Recruits from later phases or the newly minted Marines about to leave Boot Camp was at the chow hall. It seemed that was where the Drill Instructors kind of showed off their Recruits to each other. Recruit platoons were reflections of their Drill Instructors. When we stood in line, furtive glances generated a pretty good view of the world around. As we gazed at Phase III Recruits or Marines, two factors jumped out at us. First, their physical appearance was overwhelmingly superior to ours. Their covers were shaped and stiff, their sleeves looked good, they wore bloused trousers and boots, their uniforms fit well, and they all appeared strong and ready for combat. In contrast our sleeves were poorly rolled; we all looked tired and weak. We rocked go fasters, buttoned every button of our blouses up to our necks, and it seemed as though an ogre had smashed a cover onto our heads. I knew I was older and stronger than most Recruits and that only a few months separated us from them, but both myself and the rest of the platoon felt inferior to them. Appearance really does matter.

———

October 24

Beginning Week 3, everyone was sick. My eyes were heavy with a mild illness, likely an upper respiratory infection and a cold, same as everyone else. The platoon had begun to turn on itself a bit as the strong became annoyed with the weak. Basically, during the games, there were a few who went just a little too slow and failed to accomplish petty objectives. The Drill Instructors used their failures to justify the continuation of the game. For instance, "Oh, so Yoo Joon doesn't want to make it back on time, get back and start over…" It was almost inevitable that at least one person would make a small mistake because there were just so many of us – probability was against us! And so, after some time playing the same game, people desperately screamed at Yoo Joon [or whoever] to move faster. At the same time however, we all realized that there were blocks of time made just to 'play'. Even if we had been perfect, we would have continued to play the game. We got whipped into a kind of frenzy during the game, wanting the repetition and strain to stop, latching onto a hope that it would end. Unfortunately, Phase I seemed to be all about games, all day, every day. That's what crushed our souls over time. To visually see what we did in Phase I for much of the day, see the link in this footnote.[38]

So far today, we had chow and two classes – medical care and physical therapy. Those who hated Boot Camp counted the days from chow to chow. Those who hung in there went from night to night, like me. The few who enjoyed Boot Camp went from Sunday to Sunday (religious services). They were sometimes called 'motarded,' individuals motivated to the point of retardation - in a loose sense of the word. As you might have realized by now, there were a number of terms in use that are no longer considered particularly sensitive or politically correct in our society.

After class we were issued our rifles; we went to the armory, inventoried the new gear, and drilled back to the house. The rifles were stored under double lock whenever they weren't necessary. Yesterday was Sunday, everyone went to church. Religious services were great. They were a respite from the grueling nature of Boot Camp and a marker of progress, a sign that a specific week had ended and only a finite number of days were left. On Sundays we had chow, broke off for church, which was followed by a long drill, and finally a field day at the house. Field day meant a thorough, deep clean of our area; every ledge was spotless, every mirror streak-free, not a piece of lint on the floor.

Recruits were only allowed to use the head to pee during the day. It may seem cruel, but it stemmed from efficiency's sake. Given a packed schedule, it would take too long to allow hundreds of Recruits to use the toilets, one at a time, and thereby delay the schedule. The rule was that Recruits could only defecate at night, after lights

---

[38] Marines, "Black Friday | Welcome to Bootcamp," *YouTube*, March 25, 2016, https://www.youtube.com/watch?v=U_PeamfAzeo.

were out. Drill Instructors did their best to enforce the rule, but some Recruits snuck away to do their business – which made for some amusing tales. During field day this week, Recruit Potrykus snuck into the head while other Recruits were cleaning it and pleaded with them to use it. Only a few seconds after sitting down, Staff Sergeant Esposito walked in.

"What is this!" He screamed with a big smile on his face.

"This Recruit is pooping, Sir!"

"There's no way! Get up right now!"

"Aye Sir!" Potrykus exclaimed as he started to wipe.

"Did I say wipe? Go die on the quarterdeck." As Potrykus got IT'd, Staff Sergeant Esposito grinned and said to him, "Listen, if I catch another Recruit in the head... You'll be responsible for it. You'll get IT'd with them every time." To our amusement, Potrykus spent the next couple days desperately keeping other Recruits from pooping during the day, ambushing them when they tried to furtively enter the stalls.

———

October 28

We had rifle PT for MCMAP today. We used rubber rifles in lieu of our own to both reduce the chance of damage from the sand and so that Recruits didn't accidentally shoot one another. There were too few rubber rifles for the Series. I was one of the few Recruits without a rubber rifle, so I pretended I had one by enacting the movements with an 'air rifle.' Drill Instructors Sergeant O'Brien and Staff Sergeant Esposito noticed I was doing the movements without a weapon and ran up to me, yelling as if I did.

"Decety, tighten your grip!"

"Aye Aye Sir!"

"Finger straight and off the trigger! Off the trigger, Decety!"

"Close your ejection port cover!"

"Get your rifle parallel to the deck!"

I thought it was great fun, but of course I couldn't show it. At the end of the exercise, Sergeant O'Brien came up to me and quietly said "I'm going to IT you for everything you do wrong now," which I admit was a bit terrifying to hear, but he was just trying to get inside my head.

———

Today the Squad Leaders and Guide were pulled aside for about an hour to practice drill by Drill Instructors we had never met, who apparently were more specialized in drill. Not everyone in the platoon does the same thing for every movement in drill. The procedures for the Guide and Squad Leaders were often slightly different from that of the other Recruits, requiring special periods of instruction. The Recruit leadership (Squad Leaders and Guide) were at the front of the formation and had to lead it; they had to appear not only perfect, they were also expected to help with their peers. Staff Sergeant Esposito grabbed us after we returned to ask for our help in training the rest of the platoon. I think their goal was to first show us what was expected from the top down, and then for Recruits to hold each other to that standard, and train one another in a positive feedback cycle. Unfortunately, the amount of time available to actually help our peers was not enough – in the end the effort had to come from our Instructors.

When we had class, the platoon and Company in general coughed so much it was difficult to hear the teacher or other Recruits. In class the teacher would ask for someone to yell out the slides – "I need a motivator!" There were always a set of Recruits willing to get up and yell, typically the same individuals every time. Most sounded mildly mentally incompetent. The staggered sentences they screamed came from someone willing to get up and yell something everyone could read, for absolutely no reward. Typically, the Recruits willing to scream were not the brightest, but then again someone had to do it. We drilled after class for a few hours. The level of attention and physical work was enormous. I felt absolutely drained from drilling in the sun.

By the end of the day, the excitement and happiness of being able to rest was drug-like. Recently, we had been practicing hard each day for our first big test of Boot Camp, Initial Drill. Every movement mattered and the Drill Instructors had been cracking down on every little detail - the position of the thumb, how tightly the fists were clenched, how close the elbows must be to the body. We started practicing on the parade deck this week. Some Recruits looked upon the parade deck with dread, others with star struck eyes, others with no thought whatsoever. It was absolutely massive – I'm sure a skilled pilot could land medium-sized planes there with ease. To help memorize movements in drill, we screamed ditties as a platoon related to the movement being performed.[39] On the parade deck, platoons were not allowed to yell those ditties, yet the ditties nevertheless rang in my head as we marched. Everything here revolved around endurance and the attention to the minutest detail.

Last night a large band of some ten Drill Instructors roved around the squad bays for a night inspection. They ran into Recruit Zalkowitz in the head. He was a soft-spoken Recruit who was in the head

---

[39] See for instance 0:51 in Billy Fafler, "Fox Company Close Order Drill MCRD San Diego," *YouTube*, February 8, 2015, https://www.youtube.com/watch?v=OTS1HD5OM0w.

because it was the only place with enough light to read his mail. Unfortunately for him, Drill Instructors really didn't enjoy seeing Recruits reading mail after lights were out because it meant they weren't asleep and therefore might struggle with training the next day. When he saw the band of Instructors waxing angry, Zalkowitz tried to stuff the letters into his hygiene kit. Sadly, one saw him and yelled, "Oh mail? That's tight!" The entire group began yelling at Zalkowitz, one threw his kit against the wall to try and burst the shaving cream within. Satisfied they had taught him a lesson, the group moved on to go fix the rest of the Company. As they left, Sergeant O'Brien hung back, leaned over Zalkowitz and whispered in his ear: "You know I'm going to destroy you for this right?"

"Yes Sir."

"Don't worry… I got you." Sergeant O'Brien was terrifying!

———

October 29

Today was the end of Week 3. Morale was low because we learned we would be allocated two new Drill Instructors the next day, and the games we played today were brutal. We "learned to clean our rifles" using a brush and cleaning oil; but we pretty much spent the entire time playing games; getting up and down over and over, standing sitting standing sitting standing sitting without end. Whenever the Instructor said "scrub," we yelled back, "harder, faster, Aye, Aye Sir!" If the volume started to drop at all, we would play more games.

At this point, everyone looked very unique to me – no matter that we were dressed the same and had the same haircut. Your eyes had adjusted to the commonalities, so differences became apparent. During practice for Initial Drill, our Drill Hat, Staff Sergeant

Esposito became more relaxed about punishing the entire platoon, but harder on individuals. As the time to Initial Drill closed, he cracked down on individuals other than the billet holders. Before, he would blame us, the Recruit Leadership, for the actions of others. While some improved because they didn't like to see us get IT'd for their errors, a few did not or were unable to. When someone messed up a step or a movement, he would walk up to them and say, "You screwed up, go die." He pointed behind him and the Recruit ran upstairs to get IT'd in the squad bay.

Getting IT'd and yelling had a tremendous effect on me this early in training. It made me feel terribly bitter and tired at first, then just hollow because mistakes seemed so minor or petty. Motivation plummeted for some time and it felt like nothing mattered anymore. I believe most felt this way as well. Consequently, those who received the most attention and were IT'd the most became the least motivated to work hard and thus their overall performance worsened. It's kind of messed up because those Recruits who underperformed and were IT'd were unlikely to actually improve much. That's why everyone says "Don't make yourself a target" when they give advice for Boot Camp.

———

October 30
We went for a PT session, running between stations to workout. The hardest part was the feeling of dehydration because of the scorching sun and the ditties. When we ran, we had to keep screaming "left right, left right, left" every few seconds. As a result, one was left gasping for air and performance diminished. PT for me was thus not particularly optimal. Maintaining my strength would have to be on my own time. At one point during PT I yelled at my Squad Leaders to align their squads and

from exhaustion they didn't respond. "You've got some great Squad Leaders, Guide," Staff Sergeant Sheehan said ironically. That comment really hurt their morale; we all took his opinions and words to heart.

There was this one Recruit named Yoo Joon. His movements were always slow, and he didn't scream as loud as the other Recruits. Such behavior definitely made him a target. He also had a strong Chinese accent and looked piteously in pain during any kind of PT. Staff Sergeant Sheehan noticed him struggling that day. He walked up to him saying, "Yoo Joon, look down and think about your failure."

"Aye, Sir!" Yoo Joon responded - though it was drawn out and sounded like 'Aye, Saaah.' We all joined Yoo Joon in looking at the floor so that our smiles wouldn't be apparent.

Thursday October 31.

It was Halloween, but we would have no way of knowing, nothing made this day special for us. We woke, got dressed, went to chow, played games, then went to PT - which consisted of a callisthenic course. Before leaving for PT we had to apply sunscreen. Recruit Quinn did something that Sergeant O'Brien apparently took offense to, prompting him to require the application of additional sunscreen.

"Give him more sunscreen," he said pointing to a Squad Leader.

"Keep going…" he said as the entire tube was emptied into Quinn's hands. Quinn covered his face with it, rubbing the thick mask into his skin. As he was about to leave and go outside, O'Brien caught him again.

"You need more sunscreen Quinn. I don't want you to burn, I want you to be fully protected." Quinn received another tube of sunscreen to the face. He came out looking like Casper the friendly ghost, and quickly became sandman outside, as the desert sand became plastered to his face.

After PT we drilled, went to chow and drilled again. We had a set of classes on language proficiency and additional bonus pay; the Marine Corps will pay you more if you speak any foreign languages that are in demand. I was so tired that I nearly fell asleep when I closed my eyes, even when I stood. I believed it was as much the long sickness we've all had, as it was the cumulative sleep deprivation. Almost a month in, no one thought about anything except Boot Camp. Sometimes people mentioned a food they enjoyed before getting there, but that's it. People had ingrained the model being enforced by Drill Instructors; it had already been written into their hardware. Recruits were personally offended and became angry when someone committed an infraction. Drill cadences rang in our heads continuously: echoes of "left right, your right left," or whatever cadence played nonstop on a loop without end. Even in dreams I heard the cadences, orders, and shouts. Had Dante been inducted into the Marine Corps, he probably would have added Boot Camp as the 10th circle of hell to his list in *Divine Comedy*.

November 2

Yesterday we practiced on the parade deck. On a column left I must have done something wrong and went too far to the right. Senior Drill Instructor Staff Sergeant Sheehan asked why that had happened as we continued to march. I told him I didn't know. "That's a stupid answer," he told me. I was trying to think of a better response when he freaked out and screamed, "Oh, Guide doesn't want to respond!" He ran us off the parade deck and IT'd the entire platoon. Then he came up to me and said, "I don't care where you're from or where you've been but if you want a career at all you have to stop behaving like an Officer and start acting like you're Enlisted. You embarrassed the platoon." That really depressed me all day, in part because I wasn't sure how to interpret his critique. I tried asking Staff Sergeant Esposito later on for some advice and he said I was too technical and needed to "get down on their level." I didn't really understand that either though. The platoon saw, I think, that I was distressed, and a large number of people asked me if I was alright which was heartening.

We had a class to receive our record file. After that, a unit of fourteen of us (the original fourteen) went to get our ID cards issued. It took forever. We walked back to the house on our own. Afterwards, we had a mock Initial Drill and a CFT. The CFT is the combat fitness test, which is a physical test in boots and cammies. It tests one's strength and endurance over three events. People were very nervous about both the drill and the CFT. Though every specific moment and event was seemed slow while we did it, the day "passed quickly..." then the week. I opened my eyes and it was already Sunday!

Boot Camp is about mastering the very basics of being a Marine. The Marine Corps takes civilians and assumes they know nothing, have no life experience, and no logic. They then select a few traits and skills that matter or fit with the expectations and requirements of a basic Marine and instruct these civilians to perform them over and over so that it becomes ingrained into who they are. These traits and skills were predominantly making a bed, shaving, running, common Marine Corps knowledge (through prac), drill, how to help mitigate combat induced life-threatening wounds (prac ap), and – above all - rapid response to simple orders. The Marine Corps is thus certain that after three months, the entire platoon would be proficient basic Marines in those few skills. Everything else was tangential. Leadership, critical thinking, initiative, creativity... any such individualistic trait was quashed. I think the reason is that Enlisted individuals going through Boot Camp will be doing a specific job for the Corps. They will be placed and trained in accordance with the needs of the Marine Corps, and it takes a long time to train them; some MOS schools (job schools) are over a year long. Conversely, most Officers can be shifted between different positions, supervising various jobs, Enlisted cannot; they have to do the job they've been trained for. So, there's really no need to try and spend time developing individuality. Worse, individuality could result in massive mistakes later on when those Marines are performing their jobs. If they try to do something different than how they have been trained, it might be catastrophic, which could result in numerous deaths. It is therefore safer to train Enlisted to be more akin to tools that respond to orders than thinking individuals. I worry that the Marine Corps may be pushing too far towards the "robot" or tool end of the spectrum; because leadership and initiative at the lowest level is imperative for adapting, winning battles, improving methods and systems, and increasing efficiency.

November 4

Looking at my calendar, it had been a month and the time had truly whizzed by. At this point the only thing I wish I had more of was letters from my loved ones. I was surprised I had not received any from some of them. The isolation of Boot Camp showed you who cared about you and who didn't.

Last night I called over Zalkowitz, one of the Squad Leaders, to my bunk. Sergeant O'Brien had told him to find a new belt for drill, for Zalkowitz' belt was scuffed, and we all needed a perfect one. Zalkowitz had told him he had none.

"Find Recruit Potrykus!" said Sergeant O'Brien.

Zalkowitz reported, "Potrykus doesn't have one Sir!"

"I don't care! Go ask Guide!"

Zalkowitz woke me, I told him all the surplus web belts we had previously bought at the PX had already been passed out. Zalkowitz went to tell Sergeant O'Brien that.

"Find one now!" he said. I called Zalkowitz over to tell him an idea and O'Brien, waiting 20 ft. away, standing still, slowly raised a Windex bottle, pointed it at us, and squeezed it. The spray harmlessly dissipated into the air and on floor. Sergeant O'Brien then slowly backed away into the Instructor hut.

That same evening, Recruit Valdez was standing on line having had his wisdom teeth pulled. Instead of responding after an oral surgery, Recruits batted their hand in front of their face so as to reduce the chance to damage the wound and slow the healing.

Sergeant O'Brien walked up to him and said: "Valdez."

"Valdez?" *Bat bat* Valdez flapped his hand in front of him.

"Valdez!" O'Brien's voice went into a high-pitched scream. *Bat bat*

"Yell Valdez!" Valdez was furiously batting his hand across his face.

"I can't hear you Valdez, yell louder!"
bat bat bat bat bat!!!!

O'Brien also flicked off Zalkowitz tonight, Zalkowitz yelled, "Received, Aye, Aye, Sir!" O'Brien stifled a smile as he walked away. This week, Sergeant O'Brien revealed himself to be hilarious.

———

November 5

Last night I had to clean and sew the flag on the guidon, the little flag every platoon carries with them that identifies the unit. I read my novel while it dried. Everyone was shocked about the book, and everyone wanted to know what I was reading. Explaining the premise of the book (*Pillars of the Earth* by Ken Follett) reminded us all just how distant normal life had become.

Today we had our first inspection. All the inspections had a name, with each successive inspection being conducted by [and named after] someone higher in the hierarchy. This was the first one we would go through, the Senior Drill Instructor inspection. People were very nervous about it. I was lucky, it wasn't bad for me. I saw a few people get yelled at and there was great clamor in the rest of the squad bay. I watched the Drill Instructors jumping around, smashing what they could, and screaming at everything and everyone. When the inspection ended though, I turned around to see utter disaster! The air was hazy with powder and dirt. No footlocker lay undisturbed. My rack had entirely disappeared! It had been somehow ripped apart and scattered across the barracks in only a few seconds. To my great amusement, the Recruit across from me had his cover placed upon his head backwards (a heinous crime).

According to fire watch, I had apparently been screaming in my sleep last night. They told me I screamed the hygiene ditty: "good evening Sir! Recruit Decety from Chicago IL, 0311 infantry regular!" And some other random orders and responses. My environment had totally consumed my subconscious.

———

November 6
We had Initial Drill today and woke up earlier than usual, perhaps around 4AM, had chow, and drilled to warm up for the test. There were two tests of our drilling abilities in Boot Camp: Initial Drill and Final Drill. Initial Drill took place near the end of First Phase, while Final Drill took place near the end of Third Phase. From what I understand, they were ways to gauge the progress of the platoon and to determine how well Drill Instructors had taught us to … well, to drill. Winning was therefore a big deal and also helped to determine which platoon in the Company would be the "honor platoon." The honor platoon was determined by the number of events won over the course of Boot Camp. It had no impact `whatsoever on our careers as Marines, but I suspect it did have an impact on the careers of the Drill Instructors.

Most Recruits cared a lot about winning honor platoon out of pride. Both Initial and Final Drill were a series of movements on the parade deck. The set of movements were determined by a random card which featured a set of orders and maneuvers. Each movement was carefully scrutinized and scored by Drill Masters - Marines whose job it was to do nothing but judge drill.[40] We ended up winning Initial Drill! I could tell the platoon was nervous because a few individuals messed up a movement called inspection arms, which we had previously perfected under less stressful conditions.

During a game after drill, one Recruit from the back of the squad bay, I think it was Valdez, got so tired of it that he ran to the quarterdeck to IT himself. I don't think the Drill Instructors even noticed that he was over

---

[40] See for instance, USA Patriotism!, "Bravo Co. Marines – Initial Drill," *YouTube*, August 24, 2018, https://www.youtube.com/watch?v=WgESkUYg_QY

there doing pushups alone because they left him in peace!

———

November 7

Though we won Initial Drill, I was fired from Guide. I had messed up a couple things that day, like bringing a belt down and not yelling 'Aye, Aye, Sir' when told to scream a ditty. I believe the actual reason was that I had not been playing the part they wanted. The position of Guide is to be something of a bully, you're supposed to constantly yell at your fellow Recruits, force them to be disciplined, and pass down any word you receive from the Instructors. I minimized the amount of disciplining whenever I could, and instead used my Squad Leaders to enforce discipline while I tried to teach and organize the platoon in the background – particularly during free time – and led by example. It worked, in that the platoon did improve significantly; but it wasn't really what the Instructors wanted, hence the previous comment by Staff Sergeant Esposito that I was too technical. At least that was my interpretation. The day was brutal after that. I was IT'd several times and had become part of the General Population (everyone who was not a Squad Leader or Guide). It was nice because I could be buddies with everyone, actually get to know them, and had tons of time to do things like eat. On the other hand, I had to play all the stupid games to the last detail – the worst being to make my rack. As Guide I had enjoyed a bit more leeway to make my rack because I had to do so alone, so I had been somewhat exempt from ripping it apart and making it again during the games in the morning.

Yesterday was Sunday. We spent Sunday playing games, then did prac. When treating a wound, we're supposed to continue talking to keep the wounded calm and present. Recruits typically said 'you're going to be okay, you're going to go home' or 'what kind of chow do you like? I like pizza' over and over. I thought it was entertaining to recite poetry (at this instant Percy Shelley's *Ozymandias*), which – in this case – coincided with the Company Commanding Officer's round about the squad bays, he heard the poem and it made him smile. Naturally I was told not to recite poetry again after that. As usual, we cleaned the squad bay for hours in the evening. I thought about my friends and softly sung Elvis Presley songs with Recruit Potrykus while mopping. Alone with my thoughts for so long, isolated in a world of stress, I look back differently on my past experiences.

Service with the Buddhists was only an hour long. I felt depressed all the way until BDR. That night, we received the right to pray in groups. The Buddhists came together and hummed "Domo Arigato" and "Squirtle Pikachu," in lieu of any actual prayer (since we of course knew none for none of us were actually Buddhists) and tried to hide our chuckling from the patrolling Instructors. I had fire-watch that night, and read for some twenty minutes, then did laundry; it was my first-time doing laundry since Boot Camp began – we had simply reused our dirty clothing up to that point. I hadn't thought about it till that time, so it was surprising to read that information when I transcribed it. We had been in Boot Camp for a long time without doing laundry. Further I distinctly recall that I was one of the first to do laundry in the platoon, it took other Recruits far longer to realize that they had the opportunity to clean their clothes on their own! To do laundry at MCRD San Diego, you had to leave the squad bay and enter a small unattached building outside. Drill Instructors did their laundry there as well. Fire-watch was supposed to do everyone's laundry and return it in the morning – but many were overwhelmed with the sheer amount of laundry; some played favorites and didn't do everyone's laundry; others simply ran out of time or were tasked to do something else by Instructors. Zalkowitz started a little "prison" business, receiving power bars in exchange for doing people's laundry. In any case, when you had fire-watch, it was a phenomenal time to ensure that your laundry got done and done properly.

———

November 8

Today we had chow then sat in a class to sign up for a retirement savings program, a military version of the 401(k) called the TSP, which stands for the Thrift Savings Plan. I drilled behind Recruit Archibald that day and for the next couple of days. It was not easy because he bobbed around and had a different gait than I, and I had become used to drilling alone. Everyone in the 'ranks' got yelled at for dumb things that were usually out of their control, such as someone else bumping into them which would cause them to be out of step. I got IT'd individually when I awoke for putting on my blouse before I put on my trousers. Life in the General Population isn't that great.

We had a swim test today. Swimming was incredibly easy. The only difficulty was being dressed in ragged cammies that became waterlogged. All the events were pretty straightforward. One consisted of removing equipment under water and retrieving a rifle, another was swimming a stretch of water in cammies. The only sketchy one, for me, involved being fully clothed, jumping in the water, then staying afloat without moving a muscle for four minutes.[41] Most passed easily, but 15

———

[41] View the following link or search "USMC Recruit Swim Test." USA Patriotism!, "Marine Corps Recruit Swim Qualification – San Diego," *YouTube*, March 9, 2012, https://www.youtube.com/watch?v=Fs7J_E95oqQ.

Recruits in my platoon did not, and had to go through the tests again! Usually it was for a small mistake, everyone passed on their second go around. Recruits are not taught to swim in Boot Camp, they are only tested to ensure they won't drown in case they end up in water (such as during an amphibious assault).

At night the Buddhists again met up in a prayer group and repeated nonsense. We mumbled through all the Pokémon we could remember. I wrote this note at 0413, on the last shift of fire-watch. I had been making friends now that I was no longer Guide, and it was nice to not be so lonely. Before writing this note, I had finished cleaning the whiskey (a storage room attached to the squad bay) and was insulted by one of the new Drill Instructors. I had had a somewhat contemptuous and insulting attitude towards them until now. But I noticed that I had started to become a bit more respectful towards them, and only because they had begun demanding it. It was so obvious that none of us held them in high regard when they first joined us – not because we didn't care about their rank, billet, or their motivation, but because they added no value to us as a platoon. They had just graduated from Drill Instructor School and never taught us anything. They both ran around during the last hour of fire-watch, harassing the fire-watch and all the Recruits beginning to wake up, telling them to scream proper greetings and power walk, or that Recruits weren't allowed to use the head yet. In short, they woke us up early and made the morning more miserable. In general, they harassed and IT'd, but didn't provide any real constructive knowledge or criticism.

About 20% of the platoon had pinkeye. Drill Instructors were constantly ordering Recruits to not touch their faces. The most salient reason was that our hands were quite dirty. It's exceedingly difficult to break such a natural reaction; when your face itches, you brush it without thinking; but touching or scratching could rapidly cause an infection like staph or cellulitis… or pinkeye. Though pinkeye is typically caused by bacteria or viruses, there seemed to be a strong correlation between our rifles and pinkeye. We believed the oil we use to clean our rifles spread it. Before the Marine Corps, I only saw a few athletes in high school catch pink eye, and everyone at school had avoided them like the plague. Here though, it spread like wildfire, and to radical effect. It almost became entertaining to see the different variations of pinkeye, one was deep red, another a light pink; the worst was the 'sheet pinkeye.' The Recruit who slept the next rack over awoke blind every day due to the thick film of pus that had streamed out of his eyes overnight and dried his eyelids shut. I don't think anyone got out of Boot Camp without having pinkeye at least once.

———

November 9

We played games all day, breaking for a class and to do prac. At the end of the day, Staff Sergeant Sheehan posted up outside of the door, at the top of the staircase, looking through the bars. Our squad bay was on the third floor, so it presented a somewhat commanding view of the terrain outside. As he waited there, taking in the nighttime air, he saw someone three stairs below scratch his face.

"You think you can scratch your face?" he yelled down from the top of the stairs.

"Good evening Sir!"

"Yeah right, sprint!"

"Aye, Aye, Sir!"

"No volume, get back," he went on messing with this random Recruit for five minutes. A little group of us had cleaned our way right beside him to hear him better, we all thought it was hilarious.

While doing Prac on the dummies today, Yoo Joon was treating the dummy for shock and mentioned cookies, Staff Sergeant Esposito heard him. "Yoo Joon, you ate a cookie?" He asked.

"Yes Sir!" Yoo Joon responded – probably without having understood the question.

"Tight. Go die."

Yoo Joon got IT'd up for what seemed like 30 minutes, which was forever... By the time he left the quarterdeck and the staircase of the squad bay, snot was running down his face and his cammies were soaked through with sweat.

––––––

November 10

We had a PFT today. I mentioned what the CFT was, the PFT is its counterpart. It stands for physical fitness tests and consists of pull-ups, crunches, and a three-mile run. We woke up and stretched for hours. The PFT went well for me, but rather unfortunately for Recruit Gardo who got 0 pull-ups! Both he and another Recruit who failed the run received a page 11 – a negative piece of paperwork in their permanent record book – for their performance. The rest of the day was pretty much wasted playing games or drilling. We also received another round of shots from medical.

In sum our platoon performed terribly on the initial PFT. As a result, they fired all the Squad Leaders and Guide (I had been hired as a Squad Leader again by then) and replaced them with the worst Recruits to humiliate and punish the platoon. None of these mediocre Recruits could march, our order totally broke apart. These were the Recruits that were constantly getting demolished and picked on by the Instructors –

almost always for a good reason. They were Recruits Gardo, Berber, Yoo Joon, and Johnson. As we drilled back to the house immediately after the personnel shift, our column turned left and a random Drill Instructor saw Gardo, who had obviously completely messed up the pivot essential to the turn. "Eyeballs," the Instructor said. Gardo didn't respond. The Drill Instructor looked at me with his jaw dropped, "Eyeballs, you."

"Click, Sir!" I responded as I continued marching past him.

"Is that the Recruit who failed PT?" he asked incredulously. "Yes Sir, Yes, it is."

They made Yoo Joon the Guide. Taking orders from Yoo Joon was hilarious since he was normally the one getting screamed at by the other Recruits and Instructors. He threatened to put everyone on fire-watch if they didn't stop talking and making noise, which we all knew he wasn't allowed to do. After the shots, we gathered outside on the little grass available in a formation. Out of the corner of my eye, I saw Staff Sergeant Esposito with a sly little grin softly say, "Hit it, Yoo Joon." Yoo Joon sprinted out in front of the platoon holding the guidon in both hands, screaming "THIS IS 25!" in imitation of Gerard Butler's (the Spartan King in the 300) "This is Sparta." On '25,' he jumped up and drove the flag into the ground with all his strength. The entire platoon lost all sense of composure and bearing, including Staff Sergeant Esposito. The more responsible among us tried to hide our smiles in our covers - for we were being observed by Sergeant O'Brien who invariably informed us immediately after Yoo Joon's performance that we were all pieces of shit.

Chapter 2: Phase II

Today we left for Camp Pendleton. On the West Coast, Phase II, which consisted of events on the range and hikes, wasn't staged at MCRD San Diego, but in Camp Pendleton, which was about 45 minutes away by bus. There were no ranges at MCRD, and the area was surrounded by buildings and an airport. In contrast Camp Pendleton was hundreds of thousands of acres, much of it free of any construction, with plenty of space for ranges.

We packed feverishly in the morning to get chow, then sat on our bags for literally hours. We did prac forever, from breakfast to lunch, then again until about 4PM. It was awful. The worst day since receiving week. Sitting on and off our bags, yelling the definition of SMEAC or whatever over and over. While everyone was obviously terribly annoyed, it did help to ensure that there were fewer weak links for prac – not that it necessarily meant that much in the first place. Finally, the vehicles arrived, a number of large trucks to carry our packs. Like a swarm of wasps, we buzzed around getting our gear and packs aboard, then ran into a set of buses on the other side of the base. We would be reunited with our gear at the destination. Driven by the desire to escape prac, we moved at great speed.

On the bus I was on, we all pretty much passed out immediately, we had been instructed to look at the ground and not make a sound. I peeked out the window to see the long-lost world whizzing on by.

The drive felt altogether too short. We arrived in Pendleton and immediately went into the dirt to get IT'd as a platoon for some reason or other. It was barren, dusty, and ugly. I felt like I had arrived on a distant and barren planet far away from Earth, one that blended the Marine Corps Recruit Depot with Mordor. We then progressed to a set of temporary squad bays. They were terribly gloomy... small with dim lights, covered in dirt and rust. We played games for the rest of the day as we unpacked. The best part was when Recruit Potrykus went up to a Drill Instructor and said, "Good evening Sir, during the move, this Recruit's pillow was discarded. What should this Recruit do?"

"Figure it out... you lost your stuff and now you want me to babysit you, right?" responded the Drill Instructor.

"No Sir!" answered Potrykus.

Two other Drill Instructors popped out of nowhere and just started slaying this guy, IT'ing him, calling him a "little baby" the entire time. It was pretty hilarious actually; ironically the episode kind of brought the morale up. Potrykus ended up stuffing some clothes into his rack to make it look like he had a pillow for squad bay inspections.

That evening, we finally left the squad bay to get chow. Recruits always looked forward to chow. It marked the progress of the day and the food was typically quite good. It was often the only source of pleasure during the day. At Camp Pendleton, we were given another reason to look forward to chow: it was the sole building with any appeal. It stood alone in the dark, it was grand, well decorated, covered in modern glass, wreathed in palm trees, and illuminated by thoughtfully placed lights. It could not have been more of a contrast to the scorched earth and dilapidated buildings that composed the rest of the Recruit compound. The chow hall was heaven!

_____

At this point, I unfortunately lost my notes for much of Phase II. They went astray during a surprise inspection after the completion of Table II. That inspection was done in order to prevent any Recruits from bringing back live ammunition to MCRD; the rifles we drilled with all day were the same ones we fired with (M16A4 Service Rifles), Instructors did not want a Recruit to lose their mind and start shooting up the base – as was depicted by the scene in Stanley Kubrick's *Full Metal Jacket* of Private Pyle in the head. The journal I had been writing in was found and discarded by Instructors who thought it contained poetry or letters home.

Rather than trying to recreate the days from memory, I decided to share the high points. I recall three distinct elements of Phase II. First, it was in Phase II that we learned basic military land navigation. We were taught in class, then sent out onto a tract of land to locate boxes of ammunition that corresponded to specific coordinates on the map using a compass. In little groups of four, we set out through the afternoon, deciphering coordinates and traveling across ravines, hills, and dried riverbeds. It was eerily quiet in this quasi desert; we were never quite sure if the Drill Instructors were perched somewhere, taking notes on which Recruits entered into genial conversations. We were given several hours to do this, under minimal supervision. The freedom and gentle pace made land navigation the highlight of the past several months.

Second, most of the hikes of Boot Camp were staged in Phase II. With about 70 pounds in our packs, we had three major hikes – some inland over hills and dusty roads, and one major one on the beach which was notoriously and unimaginatively called the beach hike.

Every step forward you sank into the ground. I was in the front the entire time and watched platoons ahead of us disintegrate as their weaker Recruits fell back, unable to keep up. During hikes, there was unfortunately nothing to look at beside the Recruit ahead. Ragged brush covering parts of the desiccated soil extended into the distance, an interminable blanket of dirty green, tan, and grey. Not only was there nothing to look at, if your attention on the guy in front lapsed, a gap arose. Instinctually you ran to close it, creating a bigger gap for the guy behind, who in turn sprinted to catch up, and so on. This is called the slinking effect, it made hikes a lot more difficult for the guys in the back who, because of the platoon formation in which the tall are in front and the short are in the rear, are also the ones with the shortest strides.

Thirdly and most importantly, a major portion of Phase II was dedicated to learning marksmanship. Every day we woke in the pitch black of night, ran to the chow hall, ate with gusto, and quickly marched back so that we might get to the range as the sun crested over the horizon. There, we had classes on formal marksmanship for Table I and Table II. The 'tables' were a set of targets on which one received points to evaluate marksmanship. We went through many classes and practiced all morning and early afternoon. Finally, we were dismissed by our PMI's and wasted the rest of the day in some way or another, usually playing games.

The specialized range Instructors were called 'PMI's. They were the first normal Marines we had met in Boot Camp. They treated us like people and let us ask them questions about pretty much everything. The lessons took all morning and usually ran so long because we would get the PMI to talk about random topics like video games or gecko breeding.

Being under a lot of physical and mental stress has actually been interesting to experience with regards to the body. For instance, since my fingernails have more or less disappeared, I now understand why they were so important: to protect the fingertips! Hair keeps you warm and protects the head from minor cuts and bug bites and provides a cushion against blows. The brow shunts sweat away from the eyes and shields the eyes slightly from the sun and from glancing strikes. I started to see my body less as an integral part of myself and more as an asset - sort of a machine I must take care of in order for it to function properly. I had to be careful to not hurt it, not because I didn't want to get hurt but because it would make it operate less effectively and efficiently. It was an interesting perspective and mindset to slip into.

I do recall several funny experiences worth sharing. While half the platoon was shooting on the range, the other half would be manning the targets in 'the pits,' named after the depression under the targets to escape from the path of the bullets.[1] Down there, we operated with a lot less supervision from the Instructors because we were so spread out, leaving room for some of the more lazy Recruits to take a break. The best place to hide were the porta-potties placed near the ends of the pits. Another strategy was to keep on moving whenever an Instructor's path led them close to you. You could tell where they were because they would be greeted as they walked. "Good morning Sir" got progressively louder until you turned around and greeted them yourself; that phenomenon reminded me of running in tall grass in the summer, when every step generated a little cloud of grasshoppers, every step a Drill Instructor takes generates a little cloud of proper greeting.

---

[1] Aiir Source Military, "USMC Rifle Range Pits," *YouTube*, October 19, 2016, https://www.youtube.com/watch?v=dZKy_63dYZs.

Most Recruits relaxed a little bit and socialized with the other Recruits around them. They knew to shut up the moment a Drill Instructor would pass through, but sometimes the Instructor snuck up on them. When Recruits were caught, it wasn't possible to IT them in the pits, so the Instructors typically had them 'play with water jugs,' which usually meant running up and down the pits with two massive water jugs, yelling, "water, water, who needs water?" and refilling the canteens of other Recruits. There was this one small guy in my platoon named Johnson who had a soft high-pitched voice and this perpetual 'lost in the woods' look on his face. He was always slow at everything, but especially so at manning the targets in the pit (which delayed the entire Company because all the targets had to be ready for the line to fire again), so the Instructors would mess with him. It got so bad that for the rest of Boot Camp, whenever an Instructor said, "Hit it, Johnson!" he would respond "The water jugs are coming, Sir!"

You were supposed to be wearing your Kevlar and your flak jacket in the pits at all times. A Recruit called Mash was caught bumping his friend's Kevlar with his own, which made a thudding or clunking sound. Sergeant Reyes saw him do it and had him bash his friend about a dozen more times. Then made him bang into the walls and targets in the pits. Mash thought it was kind of funny, and so did Sergeant Reyes. Strangely enough, it created this bond between Reyes and Mash. From then on, Reyes was always hovering around Mash and making him do ridiculous tasks. He never forgot the episode with the Kevlar. For the rest of Boot Camp, any time he said, "Hit it Mash," Mash would throw his head forward into the space in front of him and yell, "Clunk." After we had become Marines, one of us asked Sergeant Reyes why he liked messing with Mash so much. With a huge grin on his face he said, "Because Mash is a bitch!"

Now that I think about it, Sergeant Reyes really came into his own around then, for as I mentioned he was a newly minted Drill Instructor. During a PT session, Reyes noticed that Balutiu was the fastest Recruit in the platoon – actually in the entire Company. In contrast, Recruit Berber was possibly the slowest. I believe he had damaged his feet and legs early on in Boot Camp because he always duck-walked and was in constant pain just shuffling around, even more so with a pack on. Eventually, Berber ended up being kicked out of Boot Camp due to his physical shortcomings. Reyes ran after Balutiu, who had already lapped Berber some five times around the course we were running. Once he caught Balutiu he pointed at Berber, "He's faster than you Balutiu, Berber is faster than you."

"There's no way Sir!" Balutiu exclaimed.

"He's already gotten his haircut and made a PX call. He's so fast his hair has already grown back. He's that fast Balutiu, and you're slow."

"Aye, Aye, Sir!"

I had the misfortune once to be in front of Mash in line for a PX call when Sergeant Reyes was on the prowl. Naturally, we were standing 'nut to butt' to occupy as little space as possible. Though we already touched, Reyes came up to us and whispered to Mash, "Get tighter Mash." Mash edged into my back, at this point his chest was 100% into my back. "Alright now get tighter. Tighter now, tighter now." Mash tried moving forward awkwardly into my back, naturally pushing me into the guy in front, edging the entire line into the packed barbershop.

Two humorous events happened to me during Phase II. The first was during drill in the afternoon following Table 1. We were practicing a column of files, which required the Squad Leader to turn his head to scream commands at their squad. Apparently, I didn't scream loud enough because Recruit Berber vaporized into existence in front of me and yelled, "you need to drop your balls, Decety!" I looked at him quizzically, until Sergeant O'Brien strolled up to tell me, "The biggest sissy in the platoon just called you a bitch."

It was hard not to burst out laughing.

The second was the day I received the most attention in all of Boot Camp. The background for it stemmed from a few days before in the evening. Our Senior Drill Instructor had ordered us to get sweets at dinner. The next day during a hike, he told me he didn't trust Recruit Potrykus because that Recruit only grabbed sweets when he knew no one was looking – meaning he would always grab a dessert when he went to Medical, for instance (which everyone did, of course). In my head, I decided that meant I should grab a cookie the next day to prove to him that I would act the same way with or without him watching. I obviously didn't think this one through very well.

At lunch, I grabbed a piece of cake. It was basically a sponge, the type of cake you'd find at a cheap kindergarten party. In real life I wouldn't have spent a dime on it, probably wouldn't even have eaten it if it were free. Staff Sergeant Esposito, standing at the end of the chow line as usual, saw me grab it and exclaimed "What the...!" as I passed him to sit down. Panicking, I sat and wolfed down the cake as quickly as I could... "Maybe he'd forget I had it!" I thought. I barely tasted the cake. In a moment, not a speck remained. Staff Sergeant Esposito came up to me.

"Oh, and it's the first thing you ate huh? Tight. You're a real piece of shit. You know what, you're fired."

So, I was no longer a Squad Leader. Again. I hoped that would be punishment enough, but I was dead wrong. When we left the chow hall and were doing pull-ups, Staff Sergeant Esposito yelled, "How was the cake, Decety?"

"Good Sir!" I responded, not wanting to seem like a liar.

"Is that so?" He must have found my response to be facetious, for he made the entire platoon get un-bloused. This meant we had to button our collars to the top and remove our boot bands, so our trousers were no longer fastened to the top of our boots. In essence, we physically reverted to first Phase Recruits. The platoon was livid. Recruits that were normally gentle flicked me off or scowled angrily. We marched from the chow hall not by drilling normally, but by shuffling and sprinting. After we ran up to the squad bay, I was IT'd on the quarterdeck.

When we exited the squad bay, Staff Sergeant Esposito came to me to say, "Decety, go find Sergeant O'Brien and ask him if you can play with the water jugs." So, I went upstairs to look around for Sergeant O'Brien. When I told him my task, he had me run around between different squad bays and pick-up trucks looking for the notorious, massive, water jugs. Serendipitously, no water jugs were to be found anywhere! By the time O'Brien and I had returned to the squad bay, the platoon had gone over to practice drill on the parade deck, so we ran over to find them. O'Brien had me sprint towards far off objects and return to him, then throw myself hail Mary's with my canteen, which involved chucking the canteen like a football into the distance and trying to catch it. Of course, it was impossible, by the third throw the canteen exploded to O'Brien's apparent glee. When we arrived at the parade deck, I tried to join in drill, but Esposito refused to let me. O'Brien tasked me to sprint between pieces of trash and stuff them into my cargo pocket. The task itself wasn't that bad, but by that point I was dehydrated and, with no canteen, could not hope to have my thirst sated. I must have picked up pieces of trash consisting of Q-tips used to clean rifles and cigarette butts for over an hour and a half. At times, I was joined by other Recruits who had messed up a movement in drill. O'Brien interrupted my task by having me salute cars hundreds of meters away and scream the proper greeting. He asked after every salute, "Did they salute back? No? Probably couldn't hear you... be louder next time!"

Every 20 minutes or so, O'Brien tried to put me back into the platoon to drill, but Esposito adamantly refused to let me for I was "a stain on the platoon." Finally, I was told to just stand and watch. Drill Instructors Sergeant Reyes and Duarte had rejoined the platoon around then, which meant it was now feasible to IT Recruits again. One could not be IT'd with one's rifle, so my fellow Recruits had to drop their rifles off somewhere; I was that somewhere. Starting with a single rifle, I ended up carrying up to eighteen at a time as people were off getting slayed in the dirt. O'Brien decided I should also participate in the fun and told me to go join my buddies and empty my pockets. This was a common way to order Recruits to get IT'd, so that nothing sharp, like a pen, could hurt the Recruit during an exercise. The patch of dirt was far enough away that Esposito and O'Brien couldn't see details of what was happening. So, when I ran over to Sergeant Reyes, I told him I had been ordered to empty my cargo pocket, and I pointed at the one full of trash. I'm not sure if he saw through my play on words, all he said was, "Alright, empty your pocket dumbass." I ran away from him and the IT ground to find a trashcan, a trip I made sure took some 15 minutes. By the time I had returned, the IT session was over, and the Drill Instructors had become infuriated with the platoon. I watched Sergeant O'Brien hurl the guidon like a javelin repeatedly, while the entire platoon rushed to pick it up – for no Recruit likes to see their guidon on the ground (since it represents their platoon). I rejoined the platoon quietly as we drilled back to the squad bay. That evening at chow, Staff Sergeant Sheehan walked over to me. "Why were you fired as a Squad Leader?" he asked.

"For eating cake, Sir."

"That's it?"

"Yes Sir."

"Oh, I don't care about that. You're a Squad Leader again."

Everyone around me grinned, their annoyance of having to un-blouse earlier replaced by pity for me after the half day of attention I had received. That night during BDR, when they called VIPs and I showed up like usual, Staff Sergeant Esposito asked what was wrong with me, assuming I was going up there to request to go to Medical, and not as a Squad Leader.

"Nothing Sir."

"So why are you up here?"

"This Recruit is a Squad Leader, Sir."

"What? I fired you."

"This Recruit was hired again by the Senior Drill Instructor, Sir."

Staff Sergeant Esposito burst into a massive grin as he walked away. Over the course of Boot Camp, I was fired and rehired as a Squad Leader and Guide about once every two weeks. It became a bit of a game with the rest of the platoon, and I learned to treasure the one or two days of being in the general population of the platoon. By the end of Boot Camp, I was a Squad Leader, and led my squad without yelling, without being a bully – quite a rarity in Recruit training. Though Drill Instructors didn't like my style much, but it was rational and respectful, and the job assigned to my squad was always done on time and up to standards. For anyone reading this intending to go to Boot Camp, I would recommend the same style of leadership.

————

Second Phase was when the Drill Instructors began to know the individuals in the platoon a little bit more. We noticed that with the names they assigned. They came up with a variety of nicknames for Recruits. I was called Deceitful Decety once, but it never stuck. A lot of them were pretty incredible, mostly because they were usually so nonsensical.

Sergeant Reyes replaced Recruit Gamel's name with Gumbo. There was another guy named Martinez, he was tall and pudgy. The Drill Instructors found the ground under his rack littered with Jolly Rancher wrappers one morning, which he had obviously received in a package in the mail, hidden away, and eaten at night. After IT'ing him they started calling him Jolly Rancher. They called Recruit Wahlquist Dayquil, Dallatore became Delmonte. Most humorous to me was Alfani. In his own words, he didn't go by his own name in Boot Camp once. Sergeant O'Brien had posted himself at the front of the squad bay, screaming repeatedly "Alfani! Alfani!"

Alfani sprinted across the room, came to attention in front of Sergeant O'Brien, and screamed, "Good afternoon Sir! Recruit Alfani reporting as ordered!"

"Shut the hell up you, you're not Alfani," responded O'Brien.

"Aye, Aye, Sir?"

"That's Alfani," O'Brien said as he pointed at Recruit Bates. Bates took the hint and ran over to report as Alfani and was forthwith known to all the Instructors as Alfani; in turn they began calling Alfani Alfonso. Why? We have no idea.

The few notes I found from second Phase are the following. We had early chow in the dark, got main packs, hiked for a few miles, and had class about rushing positions. We did buddy rushes on a course consisting mainly of barbed wire, trenches, and small walls. The Field Instructors in charge of this so-called assault course were terribly rude and angry, especially compared to our PMI's. At night we had a class from a land navigation Corporal who was so funny. His job was to oversee, or at least brief us on, the land navigation part of training – reading a map and navigating to various locations. He had a fascinating way to say, 'you understand,' and would say it after nearly every sentence. For instance, "Are you allowed to piss on my course? Yes, you understand! Are you allowed to crap on my course? No, you will not crap on my course, you understand?!"

He also said 'you understand' to convey thoughts and emotions. For instance, our Series (follow Series) did better on Land Nav than lead Series, the other half of the Company. The Corporal got very excited about this and furtively exclaimed to us,

"Follow Series, you understand?! Lead Series is garbage, you understand?" Naturally we all thought it was the funniest thing in the world and spent the rest of Boot Camp mimicking him.

Later on, we also did an assault course at night during which they popped flares. We had a class beforehand to tell us about how to go about the course and proper protocols for fighting in a flare-lit

217

environment. It was explained that your dominant eye should be kept closed when flares were exploding so that your pupil does not contract, that way you can see and move again immediately when the flares disappear, you don't have to wait for your eyes to adjust again. At the end of the lesson, the Instructor asked rhetorically:

"Why did pirates wear eye patches?" None of us answered.

"To protect their night vision!" he exclaimed.

On the way back from the course, Staff Sergeant Sheehan turned to us as we walked and said,

"Did we really believe that pirates wore eye patches to protect their night vision?"

"No Sir!" Zalkowitz responded.

"First time I ever heard that ... Look, you'll hear some dumb shit in the Corps."[2]

The course itself was really cool. We had to go under three or four lines of barbed wire, combat glide through three trenches, more barbed wire to high step through, and various tunnels. They periodically set off flares which filled the valley with thick smoke and amber light. When we got to the end, we had a moment without supervision to peek back and talk. The Recruits were dark shadows gliding forward through the tempest; it looked like a 21st Century rendition of hell.

Near the end of Phase II, we had probably the most interesting experience in Boot Camp. Getting gassed! The whole Company set out on the most beautiful day of Boot Camp, so far. The air felt light, the sun gentle, our backs unburdened by heavy packs. We hiked for a few miles to get to the bunker. There was a brief on what it would be like and how to properly use a gas mask. We lined up and waited for our turn to go into

---

[2] I have read, since then, that the instructor might actually have been right as some pirates may have wanted to preserve an eye in darkness in case they needed to go below deck where there was a paucity of light.

the bunker where we would be exposed to the gas. Staff Sergeant Esposito came up as we were waiting and told us, "When you get in there, turn on the vacuum!" He made a sucking sound and filled up his lungs… "then scream NATION!" Everyone turned around and started talking about whether or not they would 'turn on the vacuum,' once they got in. To my surprise afterwards, I learned that many motivated Recruits indeed had!

While we waited another Drill Instructor from a different platoon noticed that there was a Recruit named Valdez in his platoon and in ours. He got really excited, put them together, and had them hold hands and skip into the gas chamber - to our amusement and to their great discomfort.

When it was my turn, I walked into the thick bunker, and lined up against the wall facing inboard. Gas was released from a sort of framed box, floating menacingly upwards.[3] A few fans were turned on, so the smoke spread around. We broke the seal, lifting the gas masks off our faces briefly. I really didn't expect it to be like it was. I almost lost my mind, couldn't breathe, my body burned, eyes afire, melting. Worse, when we put the mask back on, I couldn't clear the gas. Phlegm and spit mixed to close my airway and covered my face. I struggled to get any air in or out. The moment I had gotten myself under control, we had to remove the mask again. This time I was ready and took a deep breath. I tried to calmly let the air out and take shallow breaths, and keep my eyes tightly closed. It was great until I put the mask back on, for when I tried to clear it and took a big breath. I promptly "died" again. It felt like all control over my life had gone. I was drowning and burning at the same time. Finally, we ran out of the bunker. My

---

[3] Documentary Recordings, "USMC GAS CHAMBER!," *YouTube,* April 20, 2018, *https*://www.youtube.com/watch?v=XZ_1mO54NsU.

experience wasn't all that bad compared to some of the others. One guy's gas mask malfunctioned, and he never found respite. One guy tried to run out the door and got tackled trying to open it. Instructors outside the gas chamber were pounding on the walls when they heard us coughing, adding chaos to the hell-like environment. Worst of all was Recruit Martinez, whose soul pretty much departed. Instead of standing in his place, stamping his feet, jumping in pain, or whatnot like most of us, he acquired this blank look and ambled about, like a lost child in a grocery store without a clue.

After dropping off my mask, I ran to the nearby porta potties; I noticed a Drill Instructor had set up a prac circlet that was really a game, and it seemed like a good time to avoid that. I hid in the porta potty until the very last moment possible; even a rank porta potty was preferable to prac. We hiked back in pretty good spirits, I think most of us actually enjoyed the experience after it was over, and we had coughed up the phlegm and snot which had clogged our respiratory system for the past several weeks. Naturally, we practiced drill for the rest of the day. Looking back on it though, the gas chamber was one of the most unique, trying, and therefore best experiences of Boot Camp.

———

I restarted my notes after Phase II. Because we had won Initial Drill, we spent team week, the week between Phase II and Phase III, in Camp Pendleton rather than back in MCRD. The rest of the Company spent that week at MCRD helping out around base and practicing being interior guards. I think it was also meant to provide a bit of a respite from Phases I and II. On Sunday we went to church in the morning. I churned out three letters there, which made me feel productive. Some people become very spiritual at Boot Camp, either because they were religious in the first place and felt that God was seeing them through, or because there was really nothing else to do. Others, like me, enjoyed the time to be human and tried to nap or write letters during services.

I came back to work on my drill cammies (the cammies we would wear for Final Drill, which had to be ironed, starched, and free of any loose strands) while others were still at their respective services - religious services began and ended at different times because of personnel and spatial requirements.

While I was ironing my cammies, Recruit Potrykus walked up to the Drill Instructor hatch with a blank expression on his face. I just knew looking at him that something terrible was going to happen. When you wanted to speak with a Drill Instructor, you slapped a piece of wood next to their door (an action called slapping the Pine) and yelled, "Good morning (or afternoon/evening) gentlemen, Recruit *blank* respectfully requests permission to speak to Senior Drill Instructor Staff Sergeant Sheehan." You always asked for the Senior Drill Instructor, even if he was not present, because he was the only one who was supposed to care about your problems. In any case with his lost look Potrykus ran up, pounded the wood softly and said in a normal speaking tone, "Recruit Potrykus requests to speak to Sergeant O'Brien."

Sergeant O'Brien ran screaming out of the room and started slaying Potrykus. He then looked at me and said, "Empty your pockets, too Decety. You should've flown across the room and punched this bitch in the face the moment he opened his stupid mouth." So, I got IT'd too! Just Potrykus and me, alone in the squad bay getting IT'd.

When the rest of the platoon got back to the squad bay, we drilled, did prac, got chow, cleaned weapons, drilled again, did a PT session, then did BDR. Today was special though. Since the rest of the Company was back in MCRD San Diego, it was just us, so the Instructors thought it would be funny to have the Guide and Squad Leaders run BDR. We were assisted by Archibald, a huge Recruit.

We acted like the Drill Instructors. We yelled at them to play with war bags, do ammo can lifts, chug water – the whole shebang. One guy was a little slow, I had him point at himself and yell, "I'm special, I'm special!" We got to take out our frustrations with the others a little bit tonight. The best one to mess with was Recruit Machado. At one point, there were five of us yelling at him, throwing his stuff around. I put his cover on his head backward and yelled at him for having it on backward.

After that night, we noticed that people yelled, "Aye, Aye, Squad Leader," in their sleep the same way they normally yelled, 'Aye, Aye, Sir.' Indeed, by this time I was far from alone in screaming responses at night!

Writing about Archibald reminded me… One morning, we did a little drill and Staff Sergeant Esposito apparently had a word with Recruit Archibald. Later on, Esposito asked the platoon, "Where are my sick bay commandos?"

"Here Sir!" A few hands shot up across the squad bay from the Recruits who often went to medical services.

"Hit it, Archibald!" said Esposito.

"You sissies!" Archibald screamed. The rest of us thought it was terribly amusing. After that, whenever Staff Sergeant Esposito asked for his sick bay commandos, without prompting the whole platoon erupted saying, 'you sissies!' or, 'effing malingerers!'

That same day, we had a school circle to talk about a move for drill called sling arms. In the move, only a set of Recruits called 'stack men' needed to have rifle slings in good condition, and many slings in the platoon – from overuse – had become loose or difficult to utilize. Yoo Joon raised his hand and Staff Sergeant Esposito pointed at him with a slight grin. Yoo Joon said, "This Recruit has noticed that some slings are better than others. This Recruit suggests that the stack men be given the better slings." Everyone thought it was a good idea, and Esposito smiled and said, "I don't know what goes on in your head Yoo Joon."

Sergeants Reyes and Duarte popped up behind him and raged, "Nothing! It's empty! Full of air!" Then they IT'd Yoo Joon.

First day of team week: Monday.
We did a card for Final Drill in the bitter cold under the starry sky. At daybreak we went to chow, the Squad Leaders ate for barely 2 minutes and rushed back. When we returned, we messed around – not playing games like usual, we literally just messed around. Most of the platoon was away doing work around the base, like counting items for a supply warehouse. The Squad Leaders and Guide drilled outside for a few hours. At one point, Staff Sergeant Esposito summoned us to the duty hut and greeted us with an open box of donuts! "Help yourselves!" he said. Being first, Recruit Edvalson eagerly grabbed one. The rest of the Drill Instructors ambushed us, emerging from crannies around the room, and screamed us out of the duty hut. We were told to watch Edvalson with our mouths open as he ate the donut he had touched. They told him he had to hydrate after such an ample snack, so he drank two canteens at once, then two more! Then they realized he also had to work

off the donut; naturally the best way to do so was with IT. But after IT he needed to hydrate, so he drank again until he puked up the donut.

We found him after the ordeal, "Was it worth it?" we asked.

"Absolutely." He said. "I would do it again… 10/10."

———

Second day of team week: Tuesday.

Last night, I was screaming in my sleep and was semi awoken by Recruit Valdez, who was telling me to shut up, which was completely understandable. Apparently, in my sleep I told him if he didn't shut up, he'd be getting fire-watch. When he cursed back, I got up and violently scribbled his name on the fire-watch roster, still half asleep, and went back to bed quite pleased at my initiative. As a result, he was awoken for fire-watch an hour later and was rather upset. Our relationship was never the same!

We all got up quickly in the morning to drill. It was completely apocalyptic outside. High winds buffeted the building, carrying all the dust of Western California into the nooks of the world. Dark sky, bitter cold as we were whipped by sand, like being in an inky whirlwind. Drilling was unique as a result, but a total disaster. We had to stop and get chow. When we returned to the squad bay, we had our intermediary PFT. It was so cold our muscles all tightened up. I personally did only 20 pull-ups in lieu of my last 27. We returned for an inspection with our Captain. He inspected Recruits Johnson, Yoo Joon, Valdez, Edvalson, Potrykus, and me. I was fortunately able to answer all his questions, but the first three Recruits were pretty bad and were unable to answer any, which is probably why he decided to inspect them in the first place.

After the inspection, we had two hours of class with the Captain about sexual assault, being in a foreign country, and combat stresses. Afterwards we drilled to chow, where the Senior Drill Instructor made us all grab desert. I got cobbler… it was incredible. That night the Recruit Leadership worked on a drill movement called column of files while everyone else cleaned.

Team week was really chill. I was on a working party in the afternoon. With a small group, I went to the gear depot and counted Kevlar covers and gloves. Then we rolled mats, which took a few hours. It's strange to feel human. The group I was in came back to the squad bay to see the rest of the platoon working on the CFT, throwing rocks to simulate throwing grenades and doing sprints.

I heard Sergeant O'Brien tell one guy he was now a motorcycle. He set off doing a motorcycle noise, brum, brum, brum, while holding onto make believe handlebars. It was so funny O'Brien had to walk behind a trash can ten meters away to laugh.

At chow that night, Staff Sergeant Esposito – having by now earned the unofficial nickname of Cookie Monster in our platoon - ordered Recruit Jasso to grab a cookie. He followed Recruit Jasso to his seat and asked, "How's the cookie Jasso?"

"Good, Sir!"

"Say mmhmmmm cookie." Staff Sergeant Esposito leaned down to his ear and repeated, "mmmmmmhmmm cookie."

"Now stare at Decety. Decety, look at Jasso. Jasso now drink your milk. Smack your lips and say, ahhh." The whole table was doing all it could to keep from bursting out in laughter.

———

Third day of team week: Wednesday

We drilled all morning with Staff Sergeant Esposito, I couldn't figure out how to call on the right foot for column of files from the left. Very frustrating; I had been calling it on the wrong foot the entire time! Other than drill, we PT'd in the afternoon, and drilled again after. Recruit Johnson became known as Cheeseburger after Sergeant Duarte yelled at him, "cover the effing Cheeseburger!"

An unexpected event occurred during drill after a column left. We were practicing kick outs, which makes it visually obvious that the platoon is switching from 15 to 30-inch steps (the inches describe the amount of distance between the steps). You literally kick your foot out and lunge back slightly on that step. I noticed that Sergeant Reyes was staring at Recruit Mash and was not surprised when he ran up to him with a mischievous light in his eyes. Once we were moving normally, so there was no kick out, Reyes said, "Ears Mash. Take a big kick out now." Mash lunged slightly forward into the Recruit in front of him.

Sergeant Reyes chuckled quietly to himself.

"Now take a REALLY big kick out Mash!" Mash leaped forward flailing his arms and crashed through all the Recruits in front of him while Reyes ran away chortling.

We took time to inventory our money valuable bags. Near me, I saw Recruit Zalkowitz take out a couple tiny seashells from his bag. Staff Sergeant Esposito also saw and ran over. "What's This?"

"Seashells Sir!"

"Where did you get these seashells?"

"This Recruit took these seashells from the PT field."

"Oh, so you stole government property?"

"Yes Sir?"

All the Drill Instructors in the building somehow heard that part and coalesced upon Zalkowitz to look at the 'stolen government property.' Of course, they slayed him for the theft. Staff Sergeant Esposito didn't forget about the seashells, every time he was near Zalkowitz, he would point at some piece of trash on the ground and say, "Look, a seashell, pick it up!" Zalkowitz became Sergeant Esposito's personal cleaner and seashell collector.

We got mail yesterday, which was nice, for we hadn't received mail in weeks because the Senior Drill Instructor had to drive to MCRD to pick it up. Today was pretty enjoyable. We cleaned our rifles in the evening and saw some of Sergeant O'Brien's pictures when he was showing us uniforms and putting up a practice T-55 Day test. He stopped on a picture of a beautiful blonde woman. Looking at us with a sly squint, he said,

"You'll never experience the joys of Parris Island... I have."[4]

_____

Fourth day of team week: Thursday
It was a beautiful, perfect day, with a slight breeze. We got a full night of sleep. We did a final drill card, then had chow and took a practice T-55 Day test.

The Instructors finally noticed that Recruit Machado had been skating past much of Boot Camp and started to pick on him. In particular, they noticed he didn't really scream. Every chow since then, they had him run around yelling, "Volume, volume, where are you!?"

Recruit Martinez had a similar, though shorter experience. While we got ready to practice drill, they

_____

[4] He was referencing that there were no women in MCRD San Diego at the time, for as I mentioned at the beginning of this book, all female basic training is held on the East coast.D

227

posted Martinez up some 50 meters away and had him war cry the entire time we marched.

In the afternoon we had free time. I couldn't believe it. The command was, "do what you want, do it now, move." This was for 30-40 minutes; we all sort of looked dazed, not knowing what to do with that. I cleaned up my gear and wrote a letter, most Recruits used the head and organized their belongings. Everyone was thrilled by the free time, even though it appeared to pass at light-speed. Senior Drill Instructor Staff Sergeant Sheehan drilled us after that on the parade deck. We got something like an hour of square away time, then passed out straight away. It was possibly the nicest day in Boot Camp.

That evening, Sergeant Reyes found Recruit Mash in a formation, where everyone was very quiet, and whispered, "Hit it, Mash."

"Clunk," Mash responded softly.

"Say it in Spanish now."

"Clunk?" he said quietly.

"That's not Spanish, dumbass," he whispered.

Mash screamed at the top of his lungs, "This Recruit doesn't know how to say clunk in Spanish Sir!" Sergeant Reyes quickly ran away from the commotion he had caused, chuckling behind his campaign cover.

———

Fifth day of team week: Friday
We got dressed before lights. Went outside and did a final drill card as usual. After some prac we did a dummy grade for drill, then went to chow. We always carried either our rifles or canteens with us. Today Recruit Quinn forgot to bring his canteen, and one of the Instructors noticed.

"Grab a rock, you!" commanded Sergeant Duarte. Quinn found a canteen-sized rock. The rest of the day, Quinn carried that rock, and proceeded to lose his mind every time Sergeant Duarte came by to tell him to, "take a drink," and, "stay hydrated Recruit."

We came back to attend a uniforms class with Sergeant O'Brien, then the Squad Leaders went outside to practice drill. We were instructed to get early chow for the first time ever, before the rest of the platoon. At the wash rack outside the chow hall, we ran into the Squad Leaders and Guide from India Company who were rather caustic, and we nearly had a little fight. The chow hall was terrific, we had plenty of time and I shoved some peach cobbler into my mouth before a Drill Instructor could spot me. I felt like the bitter critic in Pixar's *Ratatouille* when he had that magical bite which brought him back to his infancy. I could almost taste the sun that had gone into the peaches.

We ended up drilling for a bit before we were re-joined by the platoon to do a workout. During PT, they had Yoo Joon war cry the entire time because he was so far in the back. The entire platoon watched him scream, "Ahhhhh!" as he was trying to catch up. Later on, they also made him carry Archibald, the biggest Recruit. He looked like a human thumbtack that was about to be driven through the earth.

Sixth day of team week: Saturday
We woke up, got dressed, did some drill, had morning chow, then the real drill started. It was brutal. We had no water and held our weapons at trail arms for what seemed like half the afternoon, going around the house and parade deck doing turns and high knees for hours.

As we marched, two Recruits running down the road said 'good evening gentlemen' to our Drill Instructors. The guys in the front of the platoon all thought it was funny to imitate the Instructors whenever there were Recruits from other companies around, usually by criticizing something they were doing. So, as usual, one of us yelled at them. In this case, "what was that, get back" [in reference to their volume]. We were positively giddy when we watched them run away, no doubt thinking the comment had come from a Drill Instructor.

We returned to the house exhausted and immediately did PT. When we finished PT, we got chow, then more awful drill, then more PT! It was a huge contrast to the peaceful day we had had on Thursday. We were all drained after that day.

Another amusing event occurred during PT. They made Yoo Joon keep the count for pushups. At one point, Sergeant Duarte interrupted the count... "Ears."

"Open Sir!" we all screamed.

"Yoo Joon counted 6 three times, get back to 0!"

We all thought it was so funny none of us were upset.

———

Seventh day of team week: Sunday

It was the last full day of our time in Pendleton. We grabbed our sea bags and war bags, then ran out to the parade deck. The little end – those in the back of the formation – put their stuff down way too fast, so the Squad Leaders spent about an hour fixing the platoon's organization. We went to religious services afterwards, where I wrote letters, as usual. At church they had hired a rap group to sing songs like O *Come, O Come Emmanuel* and *10,000 Reasons*, which were the same songs they usually played but with a twist this time. In fact, every church service featured the same set of songs, but I give them an A for trying to mix it up a bit.

We went to chow, and as usual did pull-ups afterwards. Since the end of first Phase, when he got 0 pull-ups, the Instructors had paid a lot of attention to Gardo. Every time we did pull-ups after chow, we would hear an Instructor say, "where you at Gardo?" which always made me chuckle. Worst still was his voice, which sounded like a caveman; he responded with, "Here suhhh." The platoon thought it was both hilarious and embarrassing that he sounded like he couldn't speak English. He wasn't widely respected in the platoon because of how weak he was. No Marine can respect a man unable to lift himself. Anyways, the Instructors started calling Gardo Drago for some reason, and today after two months of practice Gardo was able to do 2 pull-ups. We all saw him and one of us asked if it was Drago or Gardo that did 2 pull-ups. Someone else yelled, "It was Drago! Gardo's buff alter ego!"

Esposito loved trying to motivate Gardo by saying, "Hit it, Gardo, do it for the Drago's!" in reference to having pride for his family name.

Chapter 3: Phase III

Unlike on the way to Pendleton, the trip back to MCRD was smooth. We packed the trucks quickly, but not in a frenzy. The bus ride was pleasant. I was worried that it would be just like when we first arrived in MCRD and revert to first Phase, or when we were moved to Camp Pendleton and played games and were IT'd all day. But that didn't happen. We got off the bus somewhat leisurely, then simply set up the house with our packs and sea bags. After dropping all our stuff off, we did another drill card, continued to set up, had chow, and cleaned up the squad bay. It was nice to be back. Strangely I felt I had returned home. Little things like the large footlocker, the clean, new head, the lack of dust on all surfaces, made me feel like I was living at the very height of luxury. I had totally lost track of the days at this point though.

———

Some Instructors from the rest of the Company came by during the night and accused the platoon of having a messy whiskey, the main closet. They proceeded to toss everything onto the ground and left the whiskey in ruin. We woke up early to fix the mess. Afterwards, we marched to chow, then divided to go to either medical or back to the house. I went to medical to receive another measles shot.

While we were sitting for an informal lecture in the middle of the day, Sergeant O'Brien looked at Quinn, who was sitting, listening to the material. Quinn always looked a little tired.

"Are you seriously falling asleep Quinn?" asked Sergeant O'Brien.

"No Sir!" he responded.

"What? You need to get punched in the face, like bad."

Later on in the day, we heard Sergeant Reyes sneak up behind Johnson and say, "Hey Johnson, you want to play a game?" He paused for a moment - frozen in place.

"It's called ITeeeee!!!"

Reyes then IT'd Johnson.

We changed into PT gear and went to the clothing depot to get new uniforms issued and spent about half the day there. Recruit Zalkowitz did an impression of Arnold Schwarzenegger as a Drill Instructor which was hilarious. We had a 'sack nasty' for lunch, then returned to the house to put the uniforms away.[1] We did some drill and field day'd the house.

We were pressed for time, so the cleaning was hectic. Drill Instructors Reyes and Duarte actually joined in to align racks, which made us feel very close to them. It was unbelievable to see them do hands on work instead of yelling at us to do it. We got a ton of mail and food that night. I ate chocolate candy, Rice Krispy treats, and two Clif Bars. Naturally, the platoon's morale was very high. Though we were told to go to bed, the platoon still had to stay awake to tape up snakes, loose straps on packs that could catch on to brush and make noise. We

---

[1] A sack nasty was a brown paper bag with an orange, a sandwich, a bag of chips, two cookies, peanut butter, condiments, and sometimes an egg. We were also given a bottle of Gatorade to wash it down. Some Instructors made their Recruits mix all their food into the chip bag and eat the resultant disgusting porridge.

used electrical tape to bundle them up neatly. Kevlar, flaks, and PT armbands had to be prepared for the next day. That took forever.

Something hilarious happened at mail with Yoo Joon that night. The Senior Drill Instructor was, as usual, passing out mail. To do so, he sat on a rack while we sat on the ground around him; it reminded me of kindergarten. He would call out the name printed upon the letter. If it was yours, you would say, "Here Sir!" and he would chuck it your way. Recruits passed the letter or package around until it made its way to you. Anyways, he reached a letter and said, "Silver." A few moments passed.

"We don't have a Silver in this platoon." He put down the letter with the other mislabeled letters and was about to continue with the rest of the mail when Yoo Joon exclaimed, "Here Sir!"

With the most incredulous look on his face, Staff Sergeant Sheehan asked, "Yoo Joon your name is Silver?"

"Yes Sir!"

"Your name is Silver?"

"Yes Sir!"

"No, Yoo Joon. Your name is Yoo Joon." A few seconds of silence went by as they stared at each other.

"This Recruit sometimes goes by Silver, Sir!"

"Listen Yoo Joon, no one here is going to call you Silver alright?"

"Aye, Sir."

Thereafter the Senior Drill Instructor almost exclusively started calling Yoo Joon Silver.

Today we woke up earlier than usual, it was a hectic morning because we had to go turn in our RCO's, our rifle scopes, at the armory. Our Instructors were all, for some reason, busy or absent, so we were marched there by another Drill Instructor who noticed our platoon was rather undisciplined. Because of that at first, he was a bit more bitter than our Instructors. But he turned out to be pretty cool, he gave us a ton of time for chow, and we all got dessert! When we came back to the house, our Instructors had returned, and the temporary one had left. We did some prac stations and went to the PX. I was incredibly happy to hear them playing *Jack and Diane* there and did a little dance in the aisles where I couldn't be seen.

We then went to our revision for MCMAP, a review of the moves we had learned over Boot Camp. The test was the day after, along with the final CFT. They sped through the movements. I was surprised how few games we played and how quiet we all were – our cohesive volume and intensity seemed to have disintegrated overnight. I was slightly apprehensive for the test because some of the more advanced movements had strange names and were sometimes rather complicated. The way we got tested was by standing in front of an instructor, who called a move and we executed it. If you didn't execute the move, you lost points. Lose a certain amount of points, fail the test.

Immediately after the review, we did a bayonet assault course.[2] You progressed through it as a fire team with prearranged spots and obstacles. The fire team would assemble at an obstacle, and a set script would be carried out. At each obstacle, the designated fire team leader yelled, "Fire team prepare to rush!"

"Prepare to rush!" the team responded.

"Rush!" said the leader.

"Rushing," everyone repeated, and altogether they moved forward.

---

[2] See the following link for a synopsis, YumaSun, "Marine Corps Bayone Assault Course," *YouTube,* June 18, 2015, https://www.youtube.com/watch?v=aMV9XiWuuy0.

The obstacles consisted of going through tunnels, under barbed wire, hiding behind logs, a rope bridge, some pits filled with tires that we had to stab, and hanging tires which we ran up to and bayoneted. It would have felt like the 19th century had they not played special effects on speakers scattered throughout, blasting sounds of helicopters, exploding grenades, and rifle fire.

When we finished, we immediately went on to "Pugil Sticks III," the third and last time we would fight one another with pugil sticks as Recruits. MCMAP required a minimum amount of actual physical fighting in order to earn a belt. Thus, we all needed the combat conditioning to earn the basic tan belt. While we waited to get into the pit in which we fought, we practiced various simple moves. The Instructors started messing around a bit. They asked who was aggressive and strong, and a bunch of Recruits shuffled to the front. Staff Sergeant Esposito saw Yoo Joon trying to appear inconspicuous and with a grin immediately sent him to the front. We saw him vanish into the pit to fight some massive dude. He emerged totally dazed. We told each other he had gone savage in there and destroyed the other guy.

We wore ample protective equipment so it was unlikely anyone would actually get physically hurt. Like Staff Sergeant Esposito, Sergeant Duarte was on the prowl and noticed Recruit Quinn was trying to hide in the background. Quinn was one of the skinniest Recruits and likely viewed as the greatest wimp in our platoon because his volume was always at 1/10 and he was never particularly intense or motivated, nor was he strong. Duarte sent Quinn to the front of the line, in front of the whole Series, and had him yell, "I'm the greatest fighter in the world!" over and over.

After he had said it a couple times, Recruit Edvalson – a motivated Recruit in our platoon who

237

didn't like that Quinn had skated through Boot Camp –
decided he had heard enough and yelled, "This Recruit
wants to challenge that statement, Sir!"

Sergeant Duarte got excited, "Get to the front,
you!" Needless to say, Edvalson beat the crap out of
Quinn.[3]

Staff Sergeant Esposito then saw me and sent me
to the front. The arena was like a pit. Upon the walls were
the Captain and various Senior Drill Instructors watching
the combat below. When I entered, Staff Sergeant
Sheehan became very excited.

I started squaring up on my opponent, the tension
was mounting. Then the Captain randomly yelled, "Hey
Decety, you ever read Clausewitz?"

"Yes Sir, *On War* Sir!"

"Good shit," he responded.

The guy lunged. We kind of squabbled messily
for a bit and I tried retreating but tripped and fell –
thereby losing that bout. The Captain yelled down,
"What would Clausewitz do!" I really had no idea what
to respond to that one during the heat of the moment. The
second bout was more in my favor and shook the guy a
bit. Then on the third round, I jabbed his face and
stunned him, waiting for him to bash me since I'd
noticed that was his favorite attack. He promptly did so
and lowered both his guard and head as he followed the
blow. I side stepped his wild swing and bunted him in
the stomach, further bending his body forward – then
took my time bringing down my stick like a baseball bat
upon his head, taking him out. The onlookers were very
excited by that finishing move. I could hear the Captain
yelling, "Clausewitz!" as I ran out of the arena.

---

[3] As I write this, I realize it appears as though instructors were picking on
Brennan but the reality is that in combat one cannot know who one will
meet to fight. Having had that experience, Brennan was in fact now better
prepared for a more dangerous scenario.

We came back from the pugil sticks to the house and went down to the first floor for a class on hazing from First Sergeant Carlson. He was this incredibly buff, positively ripped Marine who turned out to be one of the greatest personalities at Boot Camp, even though we only saw him a couple times for lessons. He explained that the crucible was the only rite of passage for the Marine Corps, "You are considered part of the brotherhood after that, and no one is allowed to haze you." He asked us if anyone felt they had been hazed unfairly in Boot Camp. One Recruit raised his hand, and Carlson called on him. He said, "The Drill Instructor had this Recruit mop the ceiling at right shoulder arms for twenty minutes last week, Sir!" We all laughed.

The First Sergeant went on, "Aye where are my fat bodies at?"

"Here Sir!" a couple Recruits respond.

He pointed at one of them and asked, "How much weight you lose so far?"

"25 pounds, Sir!"

"Good shit. Aye, you got a girlfriend?"

"No Sir!"

"Boyfriend?"

"No Sir!"

The First Sergeant looked at us again, "I gotta ask that now you understand? New Marine Corps. Aye let me tell you they're everywhere. I guarantee you there are one or two Recruits in here that get down like that."

He turned around and walked away, then looked back at us and said softly, just loud enough for us to hear him, "Those one or two Recruits are probably looking at me right now."

———

In the morning we did a little drill, hit chow, then cleaned our weapons. Afterwards we went to a uniform class. I was disappointed to learn that a lot of personal awards and qualifications could only be earned by certain MOS. In other words, it was unlikely that people with jobs like mine, which involved sitting at a computer, would be able to do cool things like attend parachute school and "earn their wings." We had some other classes during the day.

Between classes we made a head call. The Drill Instructor who was hurrying us in the head told us, "If someone's taking a crap, tell him to make a hole and piss through it. I don't give a shit. You piss on him, it's your fault for not aiming right."

After we left the head, we lined up outside ready to depart, the Drill Instructor said, "Okay zero, zero"

"Freeze, freeze!" we responded.

"Did we wash our hands?"

We looked at each other and tentatively said, "No Sir."

"Bullshit I know we wash our hands. Get back in there. Mothers of America say we need to wash our hands, you understand."

We had our MCMAP test in the afternoon. It was incredibly easy. We were tested on maybe a third of the total movements we had been taught. We returned to the squad bay to clean out weapons some more. It was a relaxed atmosphere even though Sergeant O'Brien was leading the cleaning; a very far cry from first Phase weapons cleaning. Nobody forced you to clean specific parts, I did whatever I wanted with my rifle, and dry cleaned it for inspection. I used very little cleaning and lubricating oil because that would attract dust. Those who passed the MCMAP test gained the right to wear the tan belt, I think everyone passed. We had to cut the ends of it to tailor its length, so the Squad Leaders cut all the ends off, then melted the tips with lighters so the new ends wouldn't fray.

The most amusing thing that day happened during evening BDR. We noticed that Machado, one of the more physically unfit Recruits and possibly the worst at drill in the platoon was doing pushups next to his rack. Staff Sergeant Esposito also noticed and said, "Machado! Just because you decided after two and a half months to start improving yourself does not mean you'll do better on the CFT tomorrow. You won't improve overnight."

The rest of us started asking Machado every day if he was stronger now.

————

Woke up, did drill outside in the bitter cold like at camp Pendleton. We went to chow and were granted a lot of time to eat, for the CFT I presume. We came back for a briefing on the CFT, during the question part, I asked, "Good morning Sir, this Recruit requests knowledge."

"What?" responded Staff Sergeant Esposito.

"Has Machado's pushup count increased since last night?"

"Shut the hell up," he said as we all grinned.

The CFT went well, no surprises there. We came back, showered and changed into our cammies to go to a class on marriage and divorce, which tends to be a big problem in the Marine Corps. The gist was that Marines married too young without enough forethought, in an attempt to desperately cling to a piece of normal life, to leave the barracks, or because the government paid more if you have a spouse. It was followed up by a class on proper civilian attire by the First Sergeant. As always with the First Sergeant, it was entertaining.

Some highlights were, "Look, some females wear yoga pants. Not everyone can rock that. Some of you probably wear that shit," he said as he glared around for a few moments. He then pointed at a picture showing unacceptable grooming standards, a wacky haircut the Marine Corps would call overly eccentric.

"I don't know what the hell that is…"

Randomly he exclaimed, "Aye, who here wears skinny jeans? Where are my skinny jean Recruits? You should just wear yoga pants you'll get more attention that way. The booty warrior gonna find you. He see those skinny jeans he'll take your shit in a minute. Y'all don't know the booty warrior? Look that up when you get outta here!"

After the class we drilled, came back to the squad bay, and did prac. Sergeant Duarte said two funny things during that time. Recruit Potrykus had done very poorly on the rifle qualification during grass week (in Phase II). You earned a badge you were supposed to wear on formal occasions, a badge that showed how you had shot that year – the best being 'rifle expert,' the worst being 'marksman.' The badge for marksman was a square with a target inside, prompting Marines to call it a 'pizza box.' Sergeant Duarte was part of an elite shooting team in the Marine Corps, and boasted several special badges on his uniform, making him the resident shooting expert. Anyways, weeks after grass week, he randomly went up to Potrykus, and screamed, "Where's your pizza box?"

"Footlocker, Sir!"

"You should wear that everywhere with you. On every uniform. You should pin it to your forehead to let everyone know you're a shitbag!"

"Aye, Sir!"

Later on, during prac we received a brief on the final exam, the T-55 Day test. One of the steps was to go into a cubicle and put up magnets identifying specific Marine Corps knowledge like the names of uniforms.

We had been instructed what felt like infinite times to make sure all our magnets were placed correctly, and to make sure none of them had fallen. After Sergeant O'Brien repeated this advice, Sergeant Duarte popped up and said, "Ears."

"Open Sir!" we responded.

"Whoever is next to Yoo Joon for the cubicles test has to go in and throw down all his magnets."

———

We did some drill in the morning and went back to fitting to get our clothes. Recruit Edvalson, who quite liked Sergeant Reyes, ran up to him during a brief lull, "Good morning Sir; this Recruit requests knowledge!"

"What the hell do you want you?" yelled Reyes.

"Is it true you're a boxer, Sir?" I forgot what Edvalson had seen that prompted this query, but I think he noticed somewhere that Sergeant Reyes boxed for the Marine Corps.

"Get away from me!" said Sergeant Reyes - though we could see he had made a note of Edvalson's question.

We went to clothing. The uniforms were checked for a somewhat appropriate fit. We had a sack nasty for lunch outside of the building and sat in the shadow of the arches right outside. It was very peaceful. The platoon made a PX call and received haircuts before lunch. You're not supposed to spend much money there, just purchase essentials like razor blades and soap. Some Recruits, starved for freedom, spent tons of cash at the PX. Some spent $500 on extra war bags and other useless items! We ran back to the house to start unpacking our stuff from the PX before jogging back to have a class on the educational benefits provided to Marines by the federal government, such as the GI Bill. We drilled more, then received mail.

One of the outputs of the clothing depot was to submit our cammies to have our name tapes attached to them. We got them back today and wore them proudly. Before that, they did not have a family name or U.S. Marines attached. Those name tapes actually made me feel entitled to wear the uniform, to my surprise. I got annoyed at being yelled at by random Drill Instructors, which had never happened before. Now, we can't help but look down on the undisciplined and general filth of first Phase Recruits.

———

We went to chow, then did a sustainment hike as a Series, which took us around the perimeter of MCRD. Before the hike, Sergeant Reyes went up to Edvalson. "Ears Edvalson."

"Open, Sir?"

"You said you wanted to know if I was a boxer?"

"Yes Sir!"

"You want to know something?"

"Yes Sir."

"I'll never tell you; you understand."

As we walked, we got the chance to see the entire base. The part the family saw when they came to visit was beautiful, lush, and perfectly manicured. Otherwise the base was quite small. It was a bit disappointing to leave the first sight of grass in months and return to the dirt. When we got back, we went through a "PT" shower, then cleaned our weapons. I escaped to chill with the artist Recruits, Recruits who sat in the back and painted various items to commemorate the training cycle for the Drill Instructors. They essentially created some handcrafted, motivational memorabilia. When we finished cleaning, we went to chow.

At chow, Drill Instructors O'Brien and Esposito were talking to one another, chuckling softly. They stood behind Yoo Joon, who was shoveling yogurt into his mouth.

"Eyeballs Yoo Joon," one said.

Yoo Joon rapidly spun about, eyes wide, cheeks puffed out with his face covered in yogurt. The Instructors shuffled away with their faces in their hats trying not to laugh.

Discipline had broken down. Our volume had all but disappeared. The dramatic change from Phase II to III for the entire platoon was rather shocking. There were few to no games now. No games yesterday, and only a few today. At chow I saw Sergeant Duarte give the biggest honest grin to me when he watched me trying to hurry people up to dump their trash and stack their trays.

We came back from chow and went to a class on financial responsibility. I was trying to memorize parts of the poem *Lepanto* during the class, a work by G.K. Chesterton a friend had sent me in the mail. The Instructor who dismissed us after the class was huge. He looked like the offspring of an ice giant and a Southern American football player. He also said, "do it now, move" in a bizarre accent that people emulated for the next couple nights because it was just so strange.

Perhaps to make up for the lack of games over the past few days, when we returned from chow, we played games for a long time and dumped both our footlockers and war bags in the showers. People were really quiet during square away time, they hated dumping out their belongings; it took a long time to gather up personal items from the quagmire, and you could never guarantee that you'd get everything back without any damage - assuming you found everything in the first place. I had a productive fire-watch shift that night and got all my uniform maintenance done, though I did not sleep as much as I would have liked.

At one point in the afternoon before the game when we dumped our stuff, the squad bay was hectic, and everyone was feeling a little upset that the game wouldn't stop. For some reason, I had the brilliant idea to scream, "Johnson, left cargo pocket, do it now, move!" I didn't think anything would happen except maybe I'd get IT'd. But instead the Drill Instructors stopped messing with us to go laugh when they saw a few Recruits run over to Johnson and attempt to stick him into their left cargo pockets.

———

Sunday! We woke up and immediately rushed outside to do drill. I ran away, telling the Drill Instructors I was a prayer leader. It was a minor position for Recruits to do exactly what the title suggested. Prayer leaders showed up to services early to receive guidance from the various official religious authorities. As a result, they tried to go to chow before the platoon. Not religious myself, I had no desire to become a prayer leader until I found out about the early dismissal - I wish I had known earlier! It took forever for us to get in for chow, we tried all the hatches. But once in, we had a wonderful breakfast, including apple pie! At church, there was an angry dude from my Company who kept yelling at me to put my letter writing gear away. I ignored him; my friends told him off on my behalf – I noticed that using one's allies was the best way to deal with disgruntled Recruits. We came back to work on uniforms, drilled, then went to chow. We drilled all afternoon. In the evening we went back to the hallowed first building through which we had walked, the yellow footprint building, to call home briefly and ensure that graduation details had been received by our families.

Randomly that evening, Staff Sergeant Sheehan told us a story about his last cycle.

"Some seagulls were fighting and fell into the formation of my first Phase Recruits. The Recruits were losing their minds. I had to do something, so I stepped into the formation and kicked both seagulls with one swing. The birds flew through the air, spread their wings, and took off flying. My Recruits went wild, they thought it was a miracle."

———

Final Drill this morning. Everyone felt very nervous.[4] I wasn't sure if it went well. We did some prac after, got chow, then a practice test. Sergeant O'Brien told us not to get nervous during the test, "Kids lose their minds during the test and when they pick up the weapon, they put the muzzle in their shoulder instead of the butt stock."

We had a brief PT – really easy, then returned and did a few more exercises in the house. We played games for the first time in a while. Found out we had lost Final Drill by one point, for one guy had messed up our columns, making the platoon look terrible. It was a shame because we had worked on our columns for so long. Further we had been told a week earlier that the Drill Master believed our columns to be the best he had ever seen; we all thought we had that event in the bag, as it were. Recruit Johnson was singled out by Staff Sergeant Sheehan as the one to have messed it up badly. He immediately became the single most hated man in the platoon. People were very upset, and morale was terribly

[4] AiirSource Military, "Final Drill – Marine Corps Boot Camp," *YouTube*, November 30, 2016, https://www.youtube.com/watch?v=wNV4SaiIaQ&t=170s.

low. Our chance of being honor platoon appeared gone. Undaunted, the Drill Instructors immediately pivoted to working the platoon to win the final PFT and T-55 Day test. We came back to the squad bay and did another prac test.

Two amusing events happened in the evening. We were told to get on line, and Sergeant Reyes was rooted in the location where Recruit Valdez normally stood. Hesitating about where to go and being screamed at to get on line, Valdez ran up to Sergeant Reyes, leaned over, and started sprinting in place, waving his arms and head wildly from side to side. We all stood shocked. Sergeant Reyes was so surprised he walked away without saying a thing.

Later on, during BDR, Sergeant O'Brien was screaming "Fire-watch!" over and over as he so enjoyed doing. Recruit Potrykus was the closest fire-watch sentry available, and thus had to answer the call. He ran to Sergeant O'Brien in a 'Naruto sprint' from the back of the squad bay through all of us as we grinned.

———

T-55 Day test today. We got chow really early. The Series arrived at the normal classroom for the first part of the test – the maps. There were ten questions to answer about land navigation. If you dropped anything on the floor during the test, you had to raise your hand and a Drill Instructor would tell you to pick it up. As a result, for ten minutes we heard, "pick it up, pick it up pick it up." The test itself consisted of simply using a compass to identify directions and locations on the same exact map we had been using for three months.

We then went to a set of bleachers to get a brief on the cubicles portion of the exam and practical application scenarios. The Prac Hats of the three platoons in the Series all gave their little briefs which went along the lines of "It's exactly what we've been studying for weeks, don't be nervous, take your time, etc." Unsurprisingly, they were completely correct. We walked in, facing the bulkhead. We turned around when called and put some magnets on a couple posters to answer simple Marine Corps questions. It was the easiest thing in the world. We entered the large room where Buddhist services were held and went through two sets of scenarios: assembling a rifle and treating a combat wound. The scenarios were identical to those we had been doing for months (pick up the weapon, up down left right all around clear, etc.). We were called one at a time to hit these scenarios, then did a weapons functions check.

Afterwards we got chow, did another prac test at the house and went back to the original classroom for the final written test. As we walked into the room, there were Drill Instructors conversing at the entryway. O'Brien stopped his conversation to say, "Then we have the stupid ones," and waved his finger in Recruit Zalkowitz's face; Zalkowitz scurried away smiling.

The test was ridiculously easy. Everyone agreed. The hardest part was actually to stay awake. I nodded off during almost every question and barely woke up for the next one. Recruit Potrykus had the brilliant idea of putting a sharp pencil under his chin to stab himself awake when his head dipped. Some Recruits totally passed out in the comfort and heat of the classroom. Johnson missed some 30 questions for example. Recruits Johnson, Yoo Joon, Wahlquist, and Edvalson all failed, to the enormous displeasure of Sergeant O'Brien who felt he had been "embarrassed professionally," or so he screamed as he IT'd them all profusely. They re-took the test and passed.

When we returned, we worked on our uniforms. The Recruit leadership went out to drill our equivalents at Alpha Company, to train them as peers. Their Senior Instructor gave us peanut butter Clif bars, a flavor we hadn't tasted in months after we gave their guys some tips.

After the T-55 day test we had our orders given in a classroom, all of them consisted of when we were to go to Marine Combat Training, a month-long course of additional basic training for all Marines. O'Brien passed out pens to the entire Company to sign the orders. He called me over.

"You better make sure I get my pens back Decety!".

A Drill Instructor from a different platoon came up to me as I was gathering pens.

"Decety! How are you man!" he exclaimed.

"Well, Sir!" I answered, suspicious of his motives.

"Why don't you just say good?"

"It would be unsuitable given the context, Sir!"

He stared at me for a few moments then grinned, "I like you man!" He turned, opened the door, and screamed outside, "O'Brien! I like Decety!" then walked out.

About 3 seconds passed and Sergeant O'Brien sprinted into the class, face red, veins popping. The only element missing was steam blowing out his ears. He pointed his finger right in my face and screamed, "You still suck!" and ran back out.

———

We got chow and marched over to retrieve a second set of cammies with name tapes. Then we had to repack our sea bags. The platoon went to the Marine Corps museum as I stayed outside the Company office hatch to have the security clearance interview I needed for my MOS. While I sat waiting, two Captains came out and asked me if I would do whatever it took to keep up with them during the crucible.

One asked, "Even if you're sweating blood, would you keep up?"

"Yes Sir!"

"Would you sacrifice a family member to keep up?"

"…This Recruit isn't sure that's a relevant question, Sir."

"You're right disregard that last one. You might carry the series guidon during the crucible."

After they left, I was sent away by the Chief Drill Instructor of my Series. Just as I walked into my squad bay, he ran up to tell me to return immediately. The interview took forever; I came out late and went to chow alone. The only part of the chow hall that was open was for the Drill Instructors. I had the best meal of Boot Camp that day. The base seemed abandoned. There was no one to run into. It was actually a bit eerie.

When I got back to the squad bay, I watched Yoo Joon as he walked around. He had this gait, a kind of a swagger that we called the communist walk. We all found it highly entertaining. Under my very eyes I saw him bow his head and slowly spit on the deck. I was not the only one to see, for Staff Sergeant Esposito ran over to him. Pointing at the deck, he asked,

"Is that your spit Yoo Joon?

"No Sir!"

"Bullshit!"

Yoo Joon then scrubbed the entire squad bay's floor to ensure no spit could possibly remain and, presumably, to learn that one should not spit on the deck.

———

We had the final PFT today. We went to chow, then stretched. We did a few exercises as a platoon, then were given time to use the head and stretch on our own, which inevitably devolved into small groups of eagerly chattering Recruits until the Drill Instructors thought of some other group stretches.

Staff Sergeant Esposito had Yoo Joon scream "We are 25!" Yoo Joon drew it out so it sounded more like "tweeeeenty-fiiiiiiiive." He was also doing the 'knob turning increase volume' motion with his hand as Staff Sergeant Esposito wrestled with him to take back the guidon. Pretty funny to see that little tug of war, Staff Sergeant Esposito was smiling too much to be upset. The PFT went well.

Our Series Captain came up to me before the crunches portion of the PFT to say, "Decety, this is your final interview. I will not accept failure do you understand?"

"Yes, Sir!"

"You're going to do 3,000 sit ups in two minutes."

"Aye, Sir!"

After we finished the crunches, he came up to me and asked, "did you do 3K?"

"No Sir."

"So, you failed..." and he walked away. In any case I ended up carrying the series guidon. The rest of the day was spent drilling. Sergeant Reyes annoyed me by dumping my footlocker and ripping my rack apart for no reason – despite it being heavily boot banded. Our racks, beds, had to be made perfectly. Because they had to be perfect, which took time to do, and we all valued our sleep so much, we would use boot bands (similar to rubber bands with hooks on them) to tauten the sheets and blankets and keep everything in place. If the bands were not removed when you ripped off the covers, the hooks would rip the fabric apart.

That day we were drilled by Sergeant Duarte who marched us slowly at a pace that reminded me of the French foreign legion.

———

We were supposed to spend the morning learning to rappel and doing PT with logs. Instead the Recruit Leadership went to Mike Company to help teach them drill and clean their squad bay. To my surprise, they weren't there, so after briefly cleaning I spent the time working on memorizing *Lepanto*. When we returned to our squad bay, I went to dry cleaning to find my uniforms and drop off my dress blue coat in case I needed it. I was unable to find my trousers though. When the platoon dropped off their uniforms at dry cleaning, I had been indisposed or tasked with something else, and my gear went astray. It was a big source of stress at the time. I never found my original blue trousers and had to purchase an additional pair. In the evening we prepared for final inspection and got our gear ready to go back to Pendleton for the crucible. The crucible was the final major event of Boot Camp.

———

Inspection this morning. We cleaned the house then had a quick chow; rather than drill there, we sprinted to chow and back to give individuals more time to prepare themselves and their uniforms. We got into our service Alphas and were inspected innumerable times by each other and by our Instructors, then marched slowly at port arms. We ended up IP'ing people (IP means Irish Pennant, which stands for loose strands of fabric; to IP means to get rid of those strands) all morning with lighters next to the parade deck, waiting for our turn to be inspected.

Last night I asked Bates and Gamel to give me a high and tight in the head using a razor blade. Staff Sergeant Esposito was walking through and saw them at work, he smiled and continued on. He was waiting for me to come out. Another guy had done the same to himself about a week earlier; the Instructors called it a 'Chili bowl,' after the awkward shape of the hair, since the cut was made by amateurs. His was a total mess for it was completely uneven and totally disheveled. When I walked out of the head, Esposito ambushed me.

"Zero… how the fff did you get it faded?"

"With a razor Sir!"

"…Alright." He walked away. I was shocked he hadn't chewed me out. Bates and Gamel had hooked me up!

Then today before the inspection, Staff Sergeant Sheehan noticed the cut.

"It's better than mine! Why?!"

"This Recruit wanted to look good for inspection, Sir!"

"You just wanted to look better than me!"

"No Sir!"

We were inspected by a Captain. The Squad Leaders had to call their squad to attention as he passed through. My Alpha belt kept slipping upwards. The Captain asked me, "What was the most rewarding experience of Boot Camp." I answered, "To earn but a modicum of respect from my Instructors, Sir." We came back and changed, then stacked our stuff to be ready for Pendleton. To our disgruntled surprise, we started doing prac as we waited. Everyone was shocked and annoyed, especially since we had already finished the final test. We were just doing prac to keep busy and to restrict conversation. I escaped the prac by pretending to be an artist Recruit; they sat on the side painting and coloring, so I helped them color some wood. Finally, we boarded the buses. One of the buses for our platoon did not show up, so we waited for about an hour for this last vehicle to arrive.

We played some games with our main packs for a while before going to sleep in the old Pendleton barracks. Definitely a feeling of déjà vu. The crucible involved a lot of marching, a lot of open space, and a large number of obstacles and hills, all of which made the trip to Pendleton a necessity again.

———

In the morning Sergeant O'Brien asked us if we had any genuine questions about the Marine Corps before chow. He had become very cool. After chow we came back for a head call, then church. Chow, church, then chow again that day. That night I slept terribly. It was very cold even under the two blankets I had. I awoke from the cold six or so times. We had a relatively pleasant hike to the crucible barracks in the evening, a sort of bare hanger a few miles away.

After reaching the barracks, Staff Sergeant Sheehan told me the Company Commander was impressed by our responses. Sheehan didn't think anyone had messed up during the inspection. To see what he would say, I responded,

"Recruit Yoo Joon messed up, Sir."

"When has he not, though?" was Sheehan's answer. We stood there for a moment, then Sheehan said, "You know yesterday I caught him in the head taking a crap naked. Totally naked at like one in the morning…" I tried not to smile. Sheehan looked at me.

"I asked him what he was doing there… His response was that it was a habitual thing. Who does that!" I had to hide my face in my cover to keep myself from laughing.

When we got back to our barracks, we received the order: "Do what you want, do it now, move!" People worked out and did a few foot races and pull-up competitions. Most stood around and talked. I bumped into a friend from a different platoon, who told me about a 'frog off' that had happened that week. After yelling for some time, your voice got hoarse. If you pushed past that, you lost your voice. If you continued yelling, your voice would acquire this strange otherworldly tone that Drill Instructors often had. That voice was known as a frog voice. Some Recruits inevitably got it. These two guys from different platoons in particular had especially loud and noticeable frog voices. One Drill Instructor noticed and got really excited. He brought them together on his quarterdeck, invited all the nearby Instructors, and gleefully had them do a frog off.

———

We had morning chow, then were told to do whatever we wanted again. I slept for some eight and a half hours and wore all my gear, so I wasn't that cold during the night. During free time, Recruits Alfani and Mash were told to stand at attention a foot from one another, and to stare deep into one another's eyes the entire time because Staff Sergeant Esposito thought Mash hadn't shaved. They got Alfani because they caught him passing out a list of contact info for Snapchat for after Boot Camp. We went to church, ate a ton of cookies and watched the *Grinch Who Stole Christmas*. The Recruits in the room I was in played silly games to win candy and sang songs like *Let It Go* from Frozen. People slept under the tables and sprawled on the chairs after the excitement of the games and the sugar rush. Tomorrow would be the crucible.

———

Crucible Day 1

It was difficult to take notes during the crucible because we got very little sleep, no free time, and I had nowhere to keep any pen and paper.

We slept in a field house, woke up to the stars. The field house was nothing more than a hanger of steel and concrete. We neatly set up our sleeping gear and equipment and slept on rock, but it really wasn't that bad. I thought the worst part was being perpetually filthy as we cleaned ourselves with baby wipes. That and the stench of the place from having 80+ unshowered men sleeping within. The Company awoke abruptly and formed up into squads. We carried an assault pack containing food, a sweater, water, and some other miscellaneous gear - so the load wasn't too heavy. My squad hiked in the dark to a shack and sat for a moment. Apparently, we had walked there slowly for all the time that had been allocated to sitting and discussing whatever topic we were to discuss had been spent on the journey.

My squad was led by the Senior Drill Instructor, assisted by Sergeants Reyes and Duarte. The crucible involved going between different little events continuously, events that were variations of what we had already seen and done in Boot Camp. The hardest part was walking. We walked back and forth over hills and roads forever, a total of some 50 miles in 3 days.

The other difficulty was food. If I were to eat one MRE (Meal, Ready to Eat – the modern-day military field ration) per meal in Boot Camp – or three MRE's a day - I wouldn't feel full at all. We were only given three MRE's for all three days. MRE's were just rations, a condensed meal which sometimes was actually quite good. The Instructors made it sound very difficult to survive but it wasn't that hard for such a short period of time. People spent all their time talking about their food, after we staged our packs, everyone took out a little bit of food and ate it slowly. It's actually all people talked about – maybe because they were hungry, maybe because it was the first thing they actually had control over in the past three months. Halfway through the day we staged our packs and a navy corpsman checked our feet for damage. We were also given some watered-down Gatorade and an apple. Those crucible apples were the most delicious apples I have ever eaten in my life. It is amazing how hunger can improve the taste of everything.

Day 2

Sergeant O'Brien joined our squad today. His favorite thing to say was, "incoming!" and he said it all the time if we moved slowly. When an Instructor yelled, "incoming," we repeated it and threw ourselves to the ground to simulate an artillery or mortar strike. We waited there until the Instructor said, "all clear." Other than that aspect, there was little difference between today and yesterday, endless walking, one obstacle after another.

The best part of the day came from Recruits Christianson and Yoo Joon. They were doing buddy drags and fireman carries. Staff Sergeant Sheehan had Yoo Joon, or Silver, as he now called him, carry Christianson. The former was somewhere around 130 pounds and the latter was around 200. Sheehan came up to Christianson, "Eyeballs Christianson."

"Click, Sir"

"You're a pretty tight Marine on the crucible. Now look at Yoo Joon. Look at this pathetic thing. Do you see the weakness oozing out of his body?

"Yes, Sir!"

"Do you see how disgusting he is?" Yoo Joon was covered in snot and sweat.

"Yoo Joon you're a piece of shit Recruit."

"Aye, Sir!"

"Get back down Christianson."

"Yoo Joon," Staff Sergeant Sheehan pointed in the distance for Yoo Joon to carry Christianson.

That night they passed out an apple to eat in the morning before the final hike. It was apparently a big deal that everyone have this fruit. Sergeants Duarte and Reyes ran around, it seemed like half the night, screaming, "Who doesn't have an effing apple?" And, "If you don't have an effing apple tomorrow before the hike, it'll be your fault, but I swear if you come to me and say I don't have a effing apple I'll IT you with your pack on."

Day 3

Today was the reaper hike. We awoke early, while it was still night and dashed outside with our main packs that felt terribly heavy, especially after having gotten used to wearing only our assault packs. As we got into formation, there was a palpable tension. We hiked into a valley in deep darkness. All you could see were outlines of shadowy figures cast by the safety vehicle tailing us. We walked on a road that curved into the mountain. "See that light?" I heard an Instructor say, he was pointing at a distant red flash, far above, that might as well have been part of a distant constellation. "That's where we're going."

We reached the base of the hill from whence we had come after a few miles of hiking and set our packs down. The safety vehicle took off forward, then seemed to reach a slope. It was so dark we could only see a dark mass from which emanated two beams of light. The truck seemed to rotate perpendicularly, and began to go straight up, into the night sky. Up, up, up went the truck

259

that seemed to defy gravity, straight up in the dark. Finally, it leveled off hundreds of meters above us, looking more like a spaceship than a four-wheeler. "I suppose that's not so bad," I thought to myself, trying to gauge the steep climb... But then it climbed another hill, then another!

When it was completely gone from view, we rearranged the formation, and were told, 'attack the hill.' It seemed Staff Sergeant Sheehan and the Captain literally sprinted up the hill. I tried to follow, but ended up 20-25 meters behind, with no air remaining in my lungs. I felt it was impossible to take another step. Nevertheless, we all persisted. When I got to the top of the hill, I almost caught my breath. I took a glimpse behind me at the writhing disorganized mass of the Company struggling on the long slope; I turned and began to progress up the next hill. Staff Sergeant Sheehan and the Captain stopped at the base of the final hill and waited for the first few Recruits to arrive. There we sat waiting for the rest of the Company to assemble before the rush up the final hill, just in time to be greeted by the first sunlight of a new day.

At the top we had a ceremony during which we officially earned the title Marine. Here we were given our Eagle, Globe, and Anchor emblems. The commander also made a speech, and we were given an apple to restore our energy for the final hike back to the barracks. I heard that some people cried at this ceremony, but that didn't happen today. To me it felt like just another check to mark off the list of events to accomplish before I could leave.

As we walked back, the Captain – having noticed I enjoyed poetry, told me, "I've been working on a poem, want to hear it? Get back get back I hump with my pack. Get back get back I wish I was in my rack!"[5]

We were all exhausted when we returned to shower and eat. All I'll say is that a magical transformation took place then, and even though Boot Camp was not yet over, we were Marines and life improved significantly. I didn't want to give too many details about the crucible or reaper hike, and none for Marine week; the part of training in which Recruits have become Marines; because I wouldn't want to damage the experience for anyone thinking of going through Boot Camp.

After the crucible we had another class from the First Sergeant on healthy sexual behavior. He pointed at a picture showing cervical caps, "I have no idea what that is." On the next slide he pointed at a picture showing white plastic labeled 'vaginal sponge.' "What the hell is a vaginal sponge?" Someone tried to answer, and it clearly went over his head.

"Anyways, use condoms you understand?"

The second day of Marine week was Yoo Joon's birthday. Drill Instructors paid attention to birthdays, especially when they had a special affinity for certain Recruits. So, that night around 1 AM we saw Sergeant Reyes go to Yoo Joon's rack and murmur, "Psssst Yoo Joon... Yoo Joon, wake up, Yoo Joon!"

Half asleep, Yoo Joon mumbled, "Aye, Sir."

"Sit up, Yoo Joon," he whispered.

"Aye, Sir."

"Ears Yoo Joon... Incoming."

"Incoming, Aye, Sir!" Yoo Joon said as he fell back into bed, immediately asleep again.

---

[5] In Marine Corps terminology, a 'hump' is a hike.

"Do incoming once for each year you're old Yoo Joon." Yoo Joon did 19 crunches mumbling incoming each time.

For some reason, my interactions with Sergeant Reyes during Marine week revolved around him asking me how I felt. He would walk over to me whenever I was doing something. I'd give him the proper greeting because he was just standing there looking at me. Then he'd ask me how I felt about whatever it was I was doing. "How do you feel about the house?" or "How do you feel about your trousers?" or "We're going to recycle all this equipment; does that make sense? How do you feel about that?"

The last night of Marine week, Sergeant Reyes snuck up to the rack next to mine, which was where Johnson slept. He tapped Johnson awake, and out of habit, Johnson said, "Good evening, Sir." As a Marine, you're supposed to call Noncommissioned Officers by their rank, not Sir. So, he should have said, "Good evening, Sergeant," which of course gave Sergeant Reyes the bright idea to mess with him.

"Oh, so we're a Recruit, again are we? Tight. Ears Johnson," Reyes whispered quietly.

"Open Sergeant," Johnson mumbled back.

"No, no, it's Sir."

"Aye, Sir."

"Warbag on line, do it now, move." Johnson got out of bed and held out his warbag.

"Unzip the bag, do it now, move."

"Kill."

"Empty it out, do it now, move."

"Kill."

"Put all that crap back in there, do it now, move."

"Kill."

"Rack post the warbag, do it now, move. Get back in the rack, do it now, move. Tuck yourself under your blankets, do it now, move. Close your eyes, do it now, move."

"Kill."

"Johnson."

"Yes, Sir?"

"Go to sleep, do it now, move."

"Aye, Sir."

I watched this entire exchange happen before my very eyes and had to write it down that very moment because it was just so funny.

After graduation, we all went our separate ways to go home for about two weeks. I encountered a few of my friends during the next stage of basic training, called Marine Combat Training (MCT); but I never saw the majority of my platoon again. We were dispersed into our various MOS schools afterwards, then assigned to units all over the world. Fortunately, thanks to Alfani's bravery and sacrifice, we were able to stay connected on Snapchat.

Nathan Decety

Closing Remarks

Basic training in the Marine Corps is not intended to be pleasant or fun. It's supposed to be strenuous and push people out of their comfort zones. I expected to be pushed physically and mentally, to strain my abilities to the breaking point. But that did not happen. At best, Boot Camp was uncomfortable and depressing. We repeated the same basic tasks over and over; it was more a drawn-out initiation process, and less of a learning experience.

After graduation from Boot Camp, I expected MCT, the combat training that followed, to be what Boot Camp was not, a place where Marines would become proficient riflemen and practice leading their peers on the battlefield. Instead, it was worse than Boot Camp. Vast amounts of our days were wasted, we practiced certain actions just enough to gain familiarity with them, and hurriedly moved on to let the hundreds of other Marines have their turn.

Training was akin to a gilded age assembly line, young Marines simply passed through and received 'modifications' as they continued down the training pipeline. Above all we practiced waiting. Waiting on hilltops, waiting on roads, and waiting in barracks. No one joins the Marine Corps to go through that experience. Marines are supposed to be modern day Spartans. Warriors in ethos and practice. The ideology the Marine Corps subscribes to is simply not applied in basic training.

The result is that junior Marines commonly lack sense, have not developed critical thinking faculties, and require far more training to operate effectively in their

265

jobs and on a battlefield. They are dependent on NCOs and Officers to tell them what to do and how to do it. In turn those leaders often follow what has been prescribed to them, without thought, without trying to improve what they could. The result is a slow, ossified, and vulnerable Corps.

Improving it begins with basic training. I purposefully made the argument leading to my recommendations the more detailed aspect of Part I. That is because I am a fallible human who follows his own advice (in this case). If I can highlight what the intent is – why improving training is important and what the goal should be – other minds can build off my argument and ameliorate my suggestions to create a better program than I could ever dream of.

This book focused on the Marine Corps, but it can easily be adapted to the Army, which also seeks to implement and institutionalize maneuver warfare. Away from the military, many of the lessons from military practice and theory are equally applicable to other organizations – notably businesses. The 'great books' of military theory are constantly interpreted and applied to other organizations and facets of life.[1] If military strategies and tactics are reliant on an effective force of warriors and soldiers, then the same concepts developed in this book should also apply to other organizations.

---

[1] For instance: Tiha von Ghyczy, Bolko von Oetinger, Christopher Bassford, *Clausewitz on Strategy: Inspiration and Insights from a Master Strategist* (New York: John Wiley & Sons, Inc., 2002); Mark R. McNeilly, *Sun Tzu and the Art of Business: Six Strategic Principles for Managers* (Oxford, UK: Oxford University Press, 2012); Becky Sheetz-Runkle, *Sun Tzu for Women: The Art of War for Winning in Business* (Avon, MA: Adams Media, 2011); David W. Leppanen, *The Art of Business Warfare: Outmaneuvering your Competition with Military Tactics* (New York, Writers Club Press, 2000); Emily Goldstein, *Business is Warfare* (Scotts Valley, CA: CreateSpace Independent Publishing Platform, 2016); James D. Murphy, *Business is Combat: A Fighter Pilot's Guide to Winning in Modern Business Warfare* (New York: ReganBooks, 2000).

Perhaps Taylorism has become overused and organizations should consider adapting their training to empower their employees to take more initiative, to think critically, and to exercise their judgment.

For instance, a friend I met in the Marine Corps worked as the alcohol sales manager in a local grocery store chain. Sales were near zero when he took the position; by the time he left, sales had risen by tens of thousands of dollars. What did he do? He stopped following the prescriptions of the firm. The centralized headquarters forced all their stores to mirror their most successful store. The issue was that the successful store was located in an affluent area, but not all their shops were in rich neighborhoods. Because of their policy, all stores carried expensive bottles of wine and high-end craft brews. When I asked my friend what he did, his response was, "I just gave people what they actually wanted." In this case, that meant low-end cheap beer. While headquarters could have segmented the market and re-planned their sales strategy, that's expensive, slow, and would have to be redone again and again as demographics shifted. It's so much more efficient to provide some freedom to managers who can then experiment and provide their customers with pertinent goods. The economic growth rates of the developed world are dropping, in part because productivity is not increasing tremendously – despite advancements in technology. Ameliorating the individual's ability to think critically may provide a necessary improvement to productivity growth across the Western world.

Overall, I am grateful to have had the opportunity of becoming a Marine. Though there are some aspects of the Corps I would like to change, I am proud to be part of this fighting force. It has been an honor to serve the country and an even greater honor to serve in such a hallowed organization. Personally, I made enduring friendships, learned a great deal about the world and about myself, and greatly appreciate the opportunities being a Marine has provided me. I think basic military training would be useful to almost every citizen, and I sincerely believe many of the issues suffered by the U.S. could be alleviated by compulsory service of some kind (civil or military). As I have perhaps shown in Part II of this book, it is a radically different world compared to that which most of the population lives in. If one of the points of life is to go through various experiences that make one better and to generate fascinating memories, joining the military is a great way to do so.

<u>Works Cited</u>

AiirSource Military. "Drill Instructor Gives EPIC Speech – United States Marine Corps Recruit Training." *YouTube*. June 2, 2016. https://www.youtube.com/watch?v=-Ns2FkZNTC0.

AiirSource Military. "Final Drill – Marine Corps Boot Camp." *YouTube*. November 30, 2016. https://www.youtube.com/watch?v=wNV4SaiIaQ&t=1 70s.

AiirSource Military. "USMC Rifle Range Pits." *YouTube*. October 19, 2016. https://www.youtube.com/watch?v=dZKy_63dYZs.

Ainsworth, Katherine. "American Heroes: Sgt. Major Dan Daly, USMC." *U.S. Patriot Tactical*. January 12, 2015. https://blog.uspatriottactical.com/american-heroes-sgt-mjr-dan-daly-usmc/.

Arreguín-Toft, Ivan. "How the Weak Win Wars: A Theory of Asymmetric Conflict." *International Security*, Vol. 26, No. 1 (2001), pp. 93–128.

"An Assessment of US Military Power: U.S. Marine Corps." *Heritage.org*. October 5, 2017. https://www.heritage.org/military-strength/assessment-us-military-power/us-marine-corps.

Babcock, Philip, and Marks, Mindy. "Leisure College, USA: The Decline in Student Study Time." *American Enterprise Institute*. August 5, 2010.

Bacon, Lance. "Commandant Looks to Disruptive Thinkers to Fix Corps Problems." *Marine Corps Times*. March 4, 2016. https://www.marinecorpstimes.com/news/your-marine-corps/2016/03/04/commandant-looks-to-disruptive-thinkers-to-fix-corps-problems/.

Bailey, J.B.A. *Field Artillery and Fire Power*. Oxford, UK: Military Press Ltd, 2009.

Bartlett, Merrill, ed. *Assault from the Sea: Essays on the History of Amphibious Warfare*. Annapolis: Naval Institute Press, 1983.

Bellamy, Christopher. *Red God of War: Soviet Artillery and Rocket Forces*. London: Potomac Books, Inc., 1986.

Bergengruen, Vera. "Elite Troops Are Being Worked Too Hard and Spread too thin, Military Commander Warns." *McClatchy DC Bureau*. May 5, 2017. http://www.mcclatchydc.com/news/nation-world/national/national-security/article148682644.html.

Bob Dury, and Clavin, Tom. *The Last Stand of Fox Company*. New York: Grove Press, 2009.

Boot, Max. *War Made New: Weapons, Warriors, and the Making of the Modern World*. London: Penguin Books, 2006.

Brafman, Ori and Beckstrom, Rod. *The Starfish and the Spider*. New York: Portfolio, 2006.

Brennan, David. "Russia is Attacking U.S. Forces with Electronic Weapons in Syria Every Day, General Says." *Newsweek.* March 25, 2018. http://www.newsweek.com/russia-attacking-us-forces-electronic-weapons-syria-daily-general-says-900461.

Broadberry, Stephen and Harrison, Mark. *The Economics of World War I.* Cambridge: Cambridge University Press, 2005.

Brooks, Drew. "Soldiers Recount True Story Behind '12 Strong.' *US News.* January 27, 2018. https://www.usnews.com/news/best-states/north-carolina/articles/2018-01-27/soldiers-recount-true-story-behind-12-strong.

Browne, Ryan. "US Marine Corps Suffers Third Aviation Incident in Less than 24 Hours." *CNN Politics.* April 4, 2018. https://www.cnn.com/2018/04/04/politics/us-marine-corps-aircraft-incident/index.html.

"Budget of the U.S. Navy and the U.S. Marine Corps from fiscal year 2000 to 2017 (in billion U.S. dollars)." *Statista.* https://www.statista.com/statistics/239290/budget-of-the-us-navy-and-the-us-marine-corps/.

Bybelezer, Charles. "How Russia is Using Syria as a Military Guinea Pig." *The Jerusalem Post.* February 28, 2018. https://www.jpost.com/Middle-East/How-Russia-is-using-Syria-as-a-military-guinea-pig-543839.

Cherif, Abour; Adams, Gerald; Movahedzadeh, Farahnaz; Martyn, Margaret; Dunning, Jeremy. Why

Do Students Fail? *Faculty's Perspective, Higher Learning Commission, 2014 collection of papers.* 2014

Christofferson, Thomas Rodney and Christofferson, Michael Scott. *France During World War II: From Defeat to Liberation.* New York: Fordham University Press, 2006.

Clinedinst, Melissa, and Koranteng, Anna-Maria. 2017 State of College Admission. *National Association for College Admission Counseling.* https://www.nacacnet.org/news--publications/publications/state-of-college-admission/soca-chapter1/.

Collins, Arthur S. *Common Sense Training: A Working Philosophy for Leaders.* Novato, CA: Presidio Press, 1978.

Compiler Revolt. "ULTIMATE US DRILL INSTRUCTORS DESTROYING RECRUITS COMPILATION 2018." *YouTube.* January 9, 2018. https://www.youtube.com/watch?v=wM1LqadfuYQ&t=773s.

Cox, Matthew. *"Former Recruit Commander Pleads Guilty in Parris Island Hazing Scandal." Marine Corps Times.* March 12, 2018. https://www.military.com/daily-news/2018/03/12/former-Recruit-commander-pleads-guilty-parris-island-hazing-scandal.html

Daugherty III, Leo. *The Marine Corps and the State Department: Enduring Partners in United States Foreign Policy, 1798-2007.* Jefferson, NC: McFarland, 2009.

DevilDog, "Getting Smoked by a USMC Drill Instructor." YouTube. March 14, 2015. https://www.youtube.com/watch?v=kwQPJu228Kg.

Drake, Matt. "More than 200 ISIS militants WIPED OUT in precision Russian air strike in Syria." *Express.* August 21, 2017. https://www.express.co.uk/news/world/843778/ISIS-Syria-Fighters-200-killed-Russia-Airforce-Deir-ez-Zor-terror-terrorism-war.

Documentary Recordings. "USMC GAS CHAMBER!" *YouTube.* April 20, 2018. *https*://www.youtube.com/watch?v=XZ_1mO54NsU.

Eckstein, Megan. "Interview: CMC Neller Lays Out Path to Future U.S. Marine Corps." *USNI News.* August 9, 2016. https://news.usni.org/2016/08/09/interview-cmc-neller-lays-out-path-to-future-u-s-marine-corps.

Fafler, Billy. "Fox Company Close Order Drill MCRD San Diego." *YouTube.* February 8, 2015. https://www.youtube.com/watch?v=OTS1HD5OM0w.

Ford, Ken and J. Zaloga, Steven. *Overlord: The D-Day Landings.* Oxford, New York: Osprey Publishing, 2009.

Foskett, Michael. "The Impact of Divorce Among Marines, E-5 and Below, on Unit Operational Readiness." *USMC Command and Staff College: Marine Corps University.* December 4, 2013.

Fox, Amos. "The Russian-Ukrainian War: Understanding the Dust Clouds on the Battlefield." *Modern War Institute.* January 17, 2017.

https://mwi.usma.edu/russian-ukrainian-war-understanding-dust-clouds-battlefield/.

Fuentes, Gidget. "Neller: Future Marine Corps Could be an 'Older, More Experienced' Force." *USNI News*. February 8, 2018. https://news.usni.org/2018/02/08/neller-future-marine-corps-older-experienced-force

Galula, David. *Counterinsurgency Warfare: Theory and Practice*. Westport CT: Greenwood Publishing Group, Inc., 2006.

Gatchel, Theodore. Defense at Water's Edge: Defending Against the Modern Amphibious Assault. Annapolis: Naval Institute Press, 1996.

Gibbons-Neff, Thomas. "How a Four-Hour Battle Between Russian Mercenaries and U.S. Special Commandos Unfolded in Syria." The *New York Times*. May 24, 2018. https://www.nytimes.com/2018/05/24/world/middleeast/american-commandos-russian-mercenaries-syria.html.

Gigerenzer, Gerd. *Risk Savvy: How to Make Good Decisions*. New York: Penguin Books, 2015.

Gladwell, Malcolm. *Blink: The Power of Thinking Without Thinking*. New York, NY: Back Bay Books, 2007.

Goldfein, David. "Doctrine for Joint Special Operations 3-05." *Joint Chiefs of Staff*. July 16, 2014. http://www.jcs.mil/Portals/36/Documents/Doctrine/pubs/jp3_05.pdf.

Gray, Colin. *Another Bloody Century*. London: Weidenfeld & Nicolson, 2006.

Greg R. "USMC Obstacle Course." *YouTube*. May 5, 2013. https://www.youtube.com/watch?v=NNc7cvpd5qk.

Hammes, Thomas. *The Sling and the Stone: on War in the 21st Century*. St Paul, MN: Zenith Press, 2006.

Harrison, Mark, ed. *The Economics of World War II: Six Great Powers in International Comparison*. Cambridge: Cambridge University Press, 1998.

Hayes, James. The Evolution of Military Officer Personnel Management Policies: A Preliminary Study with Parallels from Industry. *Rand Corporation*. Santa Monica, CA: 1978. https://www.rand.org/content/dam/rand/pubs/reports/2006/R2276.pdf.

Headquarters United States Marine Corps. *FY 2016 EAS Enlisted Retention Survey Results*. July 14, 2016. https://www.hqmc.marines.mil/Portals/61/FY16%20ERS%20Final.pdf?ver=2016-07-14-141912-647.

Headquarters United States Marine Corps. *U.S. Marine Corps: Concepts & Programs 2013, America's Expeditionary Force in Readiness*, 2013.

Headquarters United States Marine Corps. *A Brief History of the Marine Corps Recruit Depot Parris Island, South Carolina 1891 – 1962*. WASHINGTON, D. C.: Historical branch, G-3 division headquarters, U. S. MARINE CORPS, 1962.

Hennigan, W.J. "The New American Way of War."
Time. November 30, 2017.
http://time.com/5042700/inside-new-american-way-of-war/.

Hoffman, Michael. "US Air Force Targets and Destroy
ISIS HQ Building Using Social media." *Military.com*.
June 3, 2015.
https://www.military.com/defensetech/2015/06/03/us-air-force-targets-and-destroys-isis-hq-building-using-social-media.

Hooker, Richard, ed. *Maneuver Warfare, an Anthology*.
Novato, CA: Presidio Press, 1993.

Horlander, MG Thomas. "Army FY 2017 Budget
Overview." *Defense Innovation Marketplace*. February
2016.
http://www.defenseinnovationmarketplace.mil/resource
s/Army%20FY%202017%20Budget%20Overview.pdf.

Ilmot, Chester. *The Struggle for Europe*. Hertfordshire:
Wordsworth Editions, 1997.

Ip, Greg. *The Little Book of Economics*. Hoboken, NJ:
John, Wiley and Sons Inc., 2013.

Johnson, Natalie. "Three-Quarters of Young Americans
Don't Qualify for Military Service." *The Washington
Free Beacon*. February 22, 2018.
http://freebeacon.com/national-security/three-quarters-young-americans-dont-qualify-military-service/.

Kaplan, Robert. *The Revenge of Geography: What the
Map Tells us About Coming Conflicts and the Battle
Against Fate*. New York: Random House, 2012.

Kavanagh, Jennifer. "Determinants of Productivity for Military Personnel, a Review of Findings on the Contribution of Experience, Training, and Aptitude to Military Performance." *RAND Corporation,* Santa Monica, CA: 1981.

Keating, T.J. "Doctrine for Joint Special Operations 3-05." *Joint Chiefs of Staff.* December 17, 2003.

Keegan, John. *The Face of Battle.* London: Penguin Books, 1976.

Kotter, John. *Leading Change.* Cambridge, MA: Harvard Business Press, 2012.

Kovach, Gretel. "Marines Recall 'Surreal' Attack at Afghan Camp." *The San Diego Tribune.* October 6, 2012. http://www.sandiegouniontribune.com/military/s dut-fighting-marines-recount-bastion-attack-2012oct06-story.html.

KravYoo Joonko, Stepan. "Here Are the new Russian Weapons Putin Just Showed Off." *Bloomberg.* March 1, 2018. https://www.bloomberg.com/news/articles/2018-03-01/putin-s-newest-nukes-the-weapons-he-showed-off-in-speech.

Lowry, S., Richard S. *The Gulf War Chronicles: A Military History of the First War with Iraq.* Lincoln, NE: iUniverse Inc., 2008.

Lupfer, Timothy. "The Dynamics of Doctrine: The Change in German Tactical Doctrine During the First World War." *Combat Studies Institute, U.S. Army Command and General Staff College.* 1981.

Mack, Andrew J.R. "Why Big Nations Lose Small Wars: The Politics of Asymmetric Conflict." *World Politics*, Vol. 27, No. 2 (1975), pp. 175–200.

Marcus, Jonathan. "Should Russia's new Armata T-14 Tanks Worry NATO?" *BBC News*. May 30, 2017. http://www.bbc.com/news/world-europe-40083641

Marines. "Black Friday | Welcome to Bootcamp." *YouTube*. March 25, 2016. https://www.youtube.com/watch?v=U_PeamfAzeo.

Mattis, James. "Summary of the 2018 National Defense Strategy of the United States of America: Sharpening the American Military's Competitive Edge." https://www.defense.gov/Portals/1/Documents/pubs/2018-National-Defense-Strategy-Summary.pdf.

McDermott, Roger. "Russia's Armed Forces Rehearse New 'Shock-Fire' Tactics." *Eurasia Daily Monitor,* Vol. 15, Issue 34 (2016).

Mizokami, Kyle. "The Marine Corps' Latest Weapon is a Quadcopter." *Popular Mechanics*. February 9, 2018. https://www.popularmechanics.com/military/a16762519/the-marine-corps-latest-weapon-is-a-quadcopter/

Naval History and Heritage Command. "US Ship Force Levels: 1886-Present." November 17. 2017. https://www.history.navy.mil/research/histories/ship-histories/us-ship-force-levels.html.

National Center for Education Statistics. "Undergraduate Retention and Graduation Rates." Last

Updated: May 2018.
https://nces.ed.gov/programs/coe/indicator_ctr.asp.

Neller, Robert. "2017 Marine Corps Commandant's Professional Reading List." Marine Corps University Research Library. May 16, 2017.

Neller, Robert. *Message to the Force 2017: "Seize the Initiative."*

Neller, Robert. *Message to the Force 2018: "Execute."*

O'Neal, John and Russett, Bruce. "The Kantian Peace: The Pacific Benefits of Democracy, Interdependence, and International Organizations 1885-1992." *Source Politics*, Vol. 52, No. 1 (1999), pp. 1-37.

Parker, Kim; Cilluffo, Anthony; and Stepler, Renee. "6 facts about the U.S. military and its changing demographics." *Pew Research Center: Fact Tank, News In Numbers.* Last Modified April 13, 2017. *http*://www.pewresearch.org/fact-tank/2017/04/13/6-facts-about-the-u-s-military-and-its-changing-demographics/

Quester, Aline; Kelley, Laura; Hiatt, Cathy; Shuford, Robert. "Marine Corps Separation Rates: What's Happened Since FY00?," *CNA Analysis and Solutions.* October 2008.

Rommel, Erwin. *Attacks*. Provo, UT: Athena Press, 1979.

Rottman, Gordon, and Dennis, Peter. *World War II Infantry Assault Tactics.* Oxford: Osprey Publishing, 2008.

Roser, Max. *Our World In Data.*
https://ourworldindata.org/.

Seck, Hope Hodge. "Marine Corps Osprey Squadron
Commander in Pacific Fired." *Marine Corps Times.*
February 1, 2018. https://www.military.com/daily-
news/2018/02/01/marine-corps-osprey-squadron-
commander-pacific-fired.html

Schmid, Jon. "The Diffusion of Military Technology,
Defense and Peace Economics." *Doi,* 2017.

Schmidt, F. L. and Hunter, J. E. "The validity and
utility of selection methods in personnel psychology:
Practical and theoretical implications of 85 years of
research findings." *Psychological Bulletin,* Vol. 124
No. 2, (1998) pp. 262-274.

Slim, William. *Defeat into Victory: Battling Japan in
Burma and India, 1942-1945.* New York: Cooper
Square Press, 2000.

Smith, Megan, and Zeigler, Sean. "Terrorism before
and after 9/11 – a more dangerous world?" *Research
and Politics,* Vol. 4, No. 4 (2017).

Smith, Thomas. "Money for American Commandos."
Human Events. April 23, 2008.
http://humanevents.com/2008/04/23/money-for-
american-commandos/.

Snyder, Thomas, ed. 120 Years of Education: A
Statistical Portrait. US Department of Education: Office
of Educational Research and Improvement, 1993.

Stilwell, Majoo. "Here's Why Most Americans Can't Join the Military." *Business Insider*, September 28, 2015. http://www.businessinsider.com/heres-why-most-americans-cant-join-the-military-2015-9.

The State of Obesity. "Obesity Rates & Trends.". Last Modified June 2018. https://stateofobesity.org/rates/.

Statement of Lieutenant General Robert S. Walsh - Deputy Commandant, Combat Development and Integration & Commanding General, Marine Corps Combat Development Command - and Brigadier General Joseph Shrader - Commander Marine Corps Systems Command - and Mr. John Garner - Program Executive Officer, Land Systems Marine Corps - Before the Subcommittee on Seapower of the Senate Armed Services Committee on Marine Corps Ground Programs. June 6, 2017. https://www.armed-services.senate.gov/imo/media/doc/Walsh-Shrader-Garner_06-06-17.pdf.

"Understand ASVAB Score." *Official ASVAB Testing Program Website.* http://official-asvab.com/understand_app.htm.

United States Marine Corps. *MCDP1: Warfighting.* North Charleston, SC: Create Space Independent Publishing Platform, 2010.

United States Marine Corps, *Small Wars Manual*, 1940.

United States Marine Corps. *Marine Corps Tactics.* New York: Cosimo, 2007.

U.S. Army Field Manual No. 3-24, Marine Corps Warfighting Publication No. 3-33.5, *The U.S. Army,*

Marine Corps Counterinsurgency Field Manual.
Chicago: University of Chicago Press, 2007.

U.S. Bureau of Labor Statistics, Civilian Labor Force
Participation Rate [CIVPART], retrieved from FRED,
Federal Reserve Bank of St. Louis;
https://fred.stlouisfed.org/series/CIVPART.

U.S. Bureau of Labor Statistics, Civilian Employment-
Population Ratio [EMRATIO], retrieved from FRED,
Federal Reserve Bank of St. Louis;
https://fred.stlouisfed.org/series/EMRATIO.

USA Patriotism! "Bravo Co. Marines – Initial Drill."
YouTube. August 24, 2018.
https://www.youtube.com/watch?v=WgESkUYg_QY.

USA Patriotism! "Marine Corps Recruit Swim
Qualification – San Diego." *YouTube.* March 9, 2012.
https://www.youtube.com/watch?v=Fs7J_E95oqQ.

US Military Videos. "Alpha Company MCMAP."
YouTube. November 5, 2016.
https://www.youtube.com/watch?v=OZ9qUPv-KXQ.

Ricks, Thomas. "An elusive command philosophy and
a different command culture." *Foreign Policy.*
September 9, 2011.
https://foreignpolicy.com/2011/09/09/an-elusive-
command-philosophy-and-a-different-command-
culture/.

Ricks, Thomas. "We're getting out of the Marines
because we wanted to be part of an elite force." *Foreign
Policy.* January 4, 2013.
https://foreignpolicy.com/2017/12/20/were-getting-out-

of-the-marines-because-we-wanted-to-be-part-of-an-elite-force-4/.

Rojstaczer, Stuart, and, Healy, Christopher. "Where A Is Ordinary: The Evolution of American College and University Grading, 1940–2009." *Teachers College Record.* 2012.

Roth, Andrew. "New Russian Stealth Fighter Spotted in Syria." *The Guardian.* February 22, 2018. https://www.theguardian.com/world/2018/feb/22/new-russian-stealth-fighter-spotted-in-syria.

Ryan, Camille, and Bauman, Kurt. Educational Attainment in the United States: 2015. U.S. Department of Commerce: Economics and Statistics Administration: U.S. Census Bureau, 2016.
Sengupta, Kim. "China and Russia are catching up with military power of US and West, say leading defense experts." *Independent.* February 14, 2018. https://www.independent.co.uk/news/world/politics/china-russia-us-military-challenge-western-allies-nato-strategy-war-military-balance-a8209771.html.

Sicard, Sarah and Keller, Jared. "The KC-130 Crash is The Latest Tragedy in the Marines Corps' Worsening Aviation Mishap Crisis." *Task & Purpose.* June 11, 2017. https://taskandpurpose.com/marine-plane-crash-data/.

Schogol, Jeff. "The Next Fight: The Commandant is Pushing the Corps to be Ready for a 'Violent, Violent Fight." Marine Corps Times. September 18, 2009, https://www.marinecorpstimes.com/news/your-marine-corps/2017/09/18/the-next-fight-the-commandant-is-pushing-the-corps-to-be-ready-for-a-violent-violent-fight/

Schwartz, Nelson D. "As Debt Rises, the Government Will Soon Spend More on Interest Than on the Military." New York Times. September 25, 2018. https://www.nytimes.com/2018/09/25/business/econom y/us-government-debt-interest.html

Shapley, Deborah. *Promise and Power, the Life and Times of Robert McNamara*. Boston: Little, Brown and Company, 1993.

Sims, David. "China Widens Lead as World's Largest Manufacturer." *Thomas*. March 14, 2013. https://news.thomasnet.com/imt/2013/03/14/china-widens-lead-as-worlds-largest-manufacturer.

Sledge, E.B. *With the Old Breed: At Peleliu and Okinawa*. New York, Ballantine Books, 1981.

Snow, Shawn. "Facing retention issues, the Corps needs to Recruit highest number of Marines in a decade." *Marine Corps Times*. November 5, 2018. https://www.marinecorpstimes.com/news/your-marine-corps/2018/11/05/faced-with-retention-issues-the-corps-needs-to-Recruit-highest-number-of-marines-in-nearly-a-decade/.

TEDx Talks. "Obesity is a National Security Issue: Lieutenant General Mark Hertling at TEDx MidAtlantic 2012." *YouTube*. December 6, 2012. https://www.youtube.com/watch?v=sWN13pKVp9s.

Tierney, John J. *Chasing Ghosts, Unconventional Warfare in American History.* Washington D.C.: Potomac Books, Inc., 2006.

Thomas2878. "Marine Corps Boot Camp - Contact Your Next of Kin Phone Call." *YouTube.* November 4, 2017. https://www.youtube.com/watch?v=LP2fpn8PRns.

Tuchman, Barbara. *The Guns of August: The Outbreak of World War I.* New York: Random House, 2009.

Twining, Merrill B. *No Bended Knee: The Battle for Guadalcanal.* New York, Random House Publishing Group, 1996.

Van Allen, Fox. "China's Newest Weapons of War." *CBS News.* April 13, 2017. https://www.cbsnews.com/pictures/chinas-newest-weapons-of-war/.

Van Creveld, Martin. *Command in War.* Cambridge, MA: Harvard University Press, 1985.

Varfolomeeva, Anna. "Signaling Strength: Russia's Real Syria Strength is Electronic Warfare Against the US." *The Defense Post.* May 1, 2018. https://thedefensepost.com/2018/05/01/russia-syria-electronic-warfare/.

Vine, David. "Where in the World is the U.S. Military?" *Politico Magazine.* July 2015. https://www.politico.com/magazine/story/2015/06/us-military-bases-around-the-world-119321.

Von Schell, Adolf. *Battle Leadership*. Battleboro, VT: Echo Point Books, 1933.

Walsh, Robert S., Shrader, Joseph, and Garner, John. "Statement on 'Marine Corps Ground Programs' before the Subcommittee on Sea-power." *Committee on Armed Services, U.S. Senate.*

Walters, Glenn. "Statement of General Glenn Walters Before the Senate Armed Services Subcommittee on Readiness – On Marine Corps Readiness." *Senate Armed Services Subcommittee on Readiness.* February 8, 2017. https://www.armed-services.senate.gov/imo/media/doc/Walters_02-08-17.pdf.

Watkins, Shanea and Sherk, James. "Who Serves in the U.S. Military? The Demographics of Enlisted Troops and Officers." *The Heritage Foundation.* August 21, 2008. https://www.heritage.org/defense/report/who-serves-the-us-military-the-demographics-Enlisted-troops-and-Officers.

West, Bing. *The Village*. New York: Pocket Books, 1973.

1200WOAI. "Marine Recruits Arrive at MCRD." *YouTube.* July 26, 2018. https://www.youtube.com/watch?v=YnKPpNArrdg.

About the Author

An author and a Marine, Nathan Decety is passionate about improving major institutions. Nathan studied History at the University of Chicago before joining the Marine Corps. In addition to serving our county, Nathan Decety works in the management consulting and financial services industry. He previously published a research article on Ancient Sparta and is excited to write for non-academic audiences. He currently lives in the Chicagoland area – despite the cold.